© 2016
Clement Ampadu
drampadu@hotmail.com

ISBN:978-1-365-10991-1
ID: 18833351
www.lulu.com

All rights reserved. No part of this publication may be produced or transmitted in any form or by any means, electronic or mechanical, including photocopying and recording, or in any information storage and retrieval system, without the prior written permission of the publisher.

# Contents

Dedication     3

**1 Fixed Point Theorems for Higher-Order Hardy-Rogers-Type Contraction in Cone Metric Spaces**     4
1.1 Brief Summary . . . . . . . . . . . . . . . . . . . . . . . . . . . . . . 4
1.2 Preliminaries . . . . . . . . . . . . . . . . . . . . . . . . . . . . . . . 4
1.3 Main Results . . . . . . . . . . . . . . . . . . . . . . . . . . . . . . . 7
1.4 Exercises . . . . . . . . . . . . . . . . . . . . . . . . . . . . . . . . . 10
1.5 References . . . . . . . . . . . . . . . . . . . . . . . . . . . . . . . . 11

**2 Best Proximity Point in Regular Cone Metric Spaces for Cyclic Higher-Order Hardy-Rogers Type Contraction**     12
2.1 Brief Summary . . . . . . . . . . . . . . . . . . . . . . . . . . . . . . 12
2.2 Preliminaries . . . . . . . . . . . . . . . . . . . . . . . . . . . . . . . 12
2.3 Main Results . . . . . . . . . . . . . . . . . . . . . . . . . . . . . . . 15
2.4 Exercises . . . . . . . . . . . . . . . . . . . . . . . . . . . . . . . . . 18
2.5 References . . . . . . . . . . . . . . . . . . . . . . . . . . . . . . . . 20

**3 Fixed Point Theorems for Ordered g-Weak Higher-Order Hardy-Rogers Type Contraction in Ordered Cone Rectangular Metric Spaces**     21
3.1 Brief Summary . . . . . . . . . . . . . . . . . . . . . . . . . . . . . . 21
3.2 Preliminaries . . . . . . . . . . . . . . . . . . . . . . . . . . . . . . . 21
3.3 Main Results . . . . . . . . . . . . . . . . . . . . . . . . . . . . . . . 25
3.4 Exercises . . . . . . . . . . . . . . . . . . . . . . . . . . . . . . . . . 28
3.5 References . . . . . . . . . . . . . . . . . . . . . . . . . . . . . . . . 30

**4 Higher-Order Hardy-Rogers Type Graphic Contraction on Cone Rectangular Metric Spaces**     31
4.1 Brief Summary . . . . . . . . . . . . . . . . . . . . . . . . . . . . . . 31
4.2 Preliminaries . . . . . . . . . . . . . . . . . . . . . . . . . . . . . . . 31
4.3 Main Results . . . . . . . . . . . . . . . . . . . . . . . . . . . . . . . 36
4.4 Exercises . . . . . . . . . . . . . . . . . . . . . . . . . . . . . . . . . 39
4.5 References . . . . . . . . . . . . . . . . . . . . . . . . . . . . . . . . 40

# Dedication

This book is dedicated to those who read it .

*Clement Ampadu*
*June, 2016*

# Chapter 1

# Fixed Point Theorems for Higher-Order Hardy-Rogers-Type Contraction in Cone Metric Spaces

## 1.1 Brief Summary

**Abstract A.1 1**

By replacing the real numbers by ordering Banach space [Huang Long-Guang and Zhang Xian, Cone metric spaces and fixed point theorems of contractive mappings, J. Math. Anal. Appl. 332 (2007) 1468–1476] the notion of cone metric spaces was introduced. In the present chapter we prove some fixed point theorems for higher-order Hardy-Rogers-type contractive mappings.

## 1.2 Preliminaries

**Notation A.1 1**

$E$ will denote a real Banach space

**Notation A.2 1**

$P$ will denote a subset of $E$

**Definition A.3 1**

[Huang Long-Guang and Zhang Xian, Cone metric spaces and fixed point theorems of contractive mappings, J. Math. Anal. Appl. 332 (2007) 1468–1476] $P$ is called a cone if and only if

(a) $P$ is closed, nonempty, and $P \neq \{0\}$

(b) $a, b \in \mathbb{R}, a, b \geq 0, x, y \in P \Rightarrow ax + by \in P$

(c) $x \in P$ and $-x \in P \Rightarrow x = 0$

**Definition A.4 1**

Given cone $P \subset E$, we define a partial ordering $\leq$ with respect to $P$ by $x \leq y$ iff $y - x \in P$

### Notation A.5 1

$x < y$ will mean $x \leq y$, but $x \neq y$, while $x \ll y$ will stand for $y - x \in int\, P$, were $int\, P$ denotes the interior of $P$

### Definition A.6 1

[Huang Long-Guang and Zhang Xian, Cone metric spaces and fixed point theorems of contractive mappings, J. Math. Anal. Appl. 332 (2007) 1468–1476] The cone $P$ is called normal if there is a number $K > 0$ such that for all $x, y \in E$, $0 \leq x \leq y$ implies $\|x\| \leq K\|y\|$. The least positive number satisfying, $0 \leq x \leq y$ implies $\|x\| \leq K\|y\|$, is called the normal constant of $P$

### Definition A.7 1

[Huang Long-Guang and Zhang Xian, Cone metric spaces and fixed point theorems of contractive mappings, J. Math. Anal. Appl. 332 (2007) 1468–1476] If $\{x_n\}$ is a sequence such that $x_1 \leq x_2 \leq \cdots \leq x_n \leq \cdots \leq y$ for some $y \in E$, and there is $x \in E$ such that $\|x_n - x\| \to 0$ as $n \to \infty$, then the cone $P$ is said to be regular.

Equivalently, the above definition can be stated in the following way

### Definition A.8 1

[Huang Long-Guang and Zhang Xian, Cone metric spaces and fixed point theorems of contractive mappings, J. Math. Anal. Appl. 332 (2007) 1468–1476] The cone $P$ is regular iff every decreasing sequence which is bounded from below is convergent.

### Example A.9 1

[Huang Long-Guang and Zhang Xian, Cone metric spaces and fixed point theorems of contractive mappings, J. Math. Anal. Appl. 332 (2007) 1468–1476] A regular cone is a normal cone

### Remark A.10 1

We assume $int\, P \neq \emptyset$

### Definition A.11 1

[Huang Long-Guang and Zhang Xian, Cone metric spaces and fixed point theorems of contractive mappings, J. Math. Anal. Appl. 332 (2007) 1468–1476] Let $X$ be a nonempty set. Suppose the mapping $d: X \times X \mapsto E$ satisfies

(a) $0 < d(x, y)$ for all $x, y \in X$ and $d(x, y) = 0$ iff $x = y$

(b) $d(x, y) = d(y, x)$ for all $x, y \in X$

(c) $d(x, y) \leq d(x, z) + d(z, y)$ for all $x, y, z \in X$

Then $d$ is called a cone metric on $X$, and $(X, d)$ is called a cone metric space

### Example A.12 1

[Huang Long-Guang and Zhang Xian, Cone metric spaces and fixed point theorems of contractive mappings, J. Math. Anal. Appl. 332 (2007) 1468–1476] Let $E - \mathbb{R}^2$, $P = \{(x, y) \in E : x, y \geq 0\} \subset \mathbb{R}^2$, $X = \mathbb{R}$, and $d : X \times X \mapsto E$ is such that $d(x, y) = (|x - y|, \alpha|x - y|)$, where $\alpha \geq 0$ is a constant. Then $(X, d)$ is a cone metric space.

### Definition A.13 1

[Huang Long-Guang and Zhang Xian, Cone metric spaces and fixed point theorems of contractive mappings, J. Math. Anal. Appl. 332 (2007) 1468–1476] Let $(X, d)$ be a cone metric space. Let $\{x_n\}$ be a sequence in $X$ and $x \in X$. If for every $c \in E$ with $0 \ll c$ there is $N$ such that for all $n > N$, $d(x_n, x) \ll c$, then $\{x_n\}$ is said to be convergent and $\{x_n\}$ converges to $x$, and $x$ is the limit of $\{x_n\}$

### Lemma A.14 1

[Huang Long-Guang and Zhang Xian, Cone metric spaces and fixed point theorems of contractive mappings, J. Math. Anal. Appl. 332 (2007) 1468–1476] Let $(X, d)$ be a cone metric space, $P$ a normal cone with normal constant $K$. Let $\{x_n\}$ be a sequence in $X$. Then $\{x_n\}$ converges to $x$ iff $d(x_n, x) \to 0$ as $n \to \infty$

### Lemma A.15 1

[Huang Long-Guang and Zhang Xian, Cone metric spaces and fixed point theorems of contractive mappings, J. Math. Anal. Appl. 332 (2007) 1468–1476] Let $(X, d)$ be a cone metric space, $P$ a normal cone with normal constant $K$. Let $\{x_n\}$ be a sequence in $X$. If $\{x_n\}$ converges to $x$ and $\{x_n\}$ converges to $y$, then $x = y$. That is the limit of $\{x_n\}$ is unique.

### Definition A.16 1

[Huang Long-Guang and Zhang Xian, Cone metric spaces and fixed point theorems of contractive mappings, J. Math. Anal. Appl. 332 (2007) 1468–1476] Let $(X, d)$ be a cone metric space. Let $\{x_n\}$ be a sequence in $X$ and $x \in X$. If for every $c \in E$ with $0 \ll c$ there is $N$ such that for all $n, m > N$, $d(x_n, x_m) \ll c$, then $\{x_n\}$ is called a Cauchy sequence.

### Definition A.17 1

[Huang Long-Guang and Zhang Xian, Cone metric spaces and fixed point theorems of contractive mappings, J. Math. Anal. Appl. 332 (2007) 1468–1476] Let $(X, d)$ be a cone metric space, if every Cauchy sequence is convergent in $X$, then $X$ is called a complete cone metric space.

### Lemma A.18 1

[Huang Long-Guang and Zhang Xian, Cone metric spaces and fixed point theorems of contractive mappings, J. Math. Anal. Appl. 332 (2007) 1468–1476] Let $(X, d)$ be a cone metric space, $\{x_n\}$ be a sequence in $X$. If $\{x_n\}$ converges to $x$, then $\{x_n\}$ is a Cauchy sequence.

### Lemma A.19 1

[Huang Long-Guang and Zhang Xian, Cone metric spaces and fixed point theorems of contractive mappings, J. Math. Anal. Appl. 332 (2007) 1468–1476] Let $(X, d)$ be a cone metric space, $P$ a normal cone with normal constant $K$. Let $\{x_n\}$ be a sequence in $X$. The $\{x_n\}$ is Cauchy iff $d(x_n, x_m) \to 0$ as $n, m \to \infty$.

### Lemma A.20 1

[Huang Long-Guang and Zhang Xian, Cone metric spaces and fixed point theorems of contractive mappings, J. Math. Anal. Appl. 332 (2007) 1468–1476] Let $(X, d)$ be a cone metric space, $P$ be a normal cone with normal constant $K$. Let $\{x_n\}$ and $\{y_n\}$ be two sequences in $X$ and $x_n \to x$, $y_n \to y$ as $n \to \infty$. Then $d(x_n, y_n) \to d(x, y)$ as $n \to \infty$

### Definition A.21 1

[Huang Long-Guang and Zhang Xian, Cone metric spaces and fixed point theorems of contractive mappings, J. Math. Anal. Appl. 332 (2007) 1468–1476] Let $(X, d)$ be a cone metric space. If for any sequence $\{x_n\}$ in $X$, there is a subsequence $\{x_{n_i}\}$ of $\{x_n\}$ such that $\{x_{n_i}\}$ is convergent in $X$. Then $X$ is called a sequentially compact cone metric space.

## 1.3 Main Results

### Theorem A.1 1

Let $(X, d)$ be a complete cone metric space, $P$ be a normal cone with normal constant $K$. Suppose the mapping $T : X \mapsto X$ satisfies the contractive condition, $d(T^r x, T^r y) \leq Z\beta^r[d(x, Tx) + d(y, Ty) + d(x, Ty) + d(y, Tx) + d(x, y)]$, for all $x, y \in X$, where $Z \geq 1$ and $\beta \in [0, \frac{1}{5})$ come from Proposition 1.3 [Ampadu, Clement(2016):Higher Order Hardy-Rogers Contraction Mapping Theorem. Unpublished]. Then $T$ has a unique fixed point in $X$, and for any $x \in X$, the iterative sequence $\{T^{rn}x\}$ converges to the fixed point.

### Proof of Theorem A.1 1

Choose $x_0 \in X$. Set $x_1 = T^r x_0$, $x_2 = T^r x_1 = T^{2r} x_0, \cdots, x_{n+1} = T^r x_n = T^{r(n+1)} x_0, \cdots$, then we have,

$$\begin{aligned}
d(x_{n+1}, x_n) &= d(T^r x_n, T^r x_{n-1}) \\
&\leq Z\beta^r[d(x_n, Tx_n) + d(x_{n-1}, Tx_{n-1}) + d(x_n, Tx_{n-1}) \\
&\quad + d(x_{n-1}, Tx_n) + d(x_n, x_{n-1})] \\
&= Z\beta^r[d(x_n, x_{n+1}) + d(x_{n-1}, x_n) + d(x_{n-1}, x_{n+1}) + d(x_n, x_{n-1})] \\
&\leq Z\beta^r[2d(x_n, x_{n+1}) + 3d(x_{n-1}, x_n)]
\end{aligned}$$

From the above it follows that $d(x_{n+1}, x_n) \leq kd(x_n, x_{n-1})$, where $k := \frac{3Z\beta^r}{1-2Z\beta^r}$. Now we observe that

$$\begin{aligned}
d(x_{n+1}, x_n) &= d(T^r x_n, T^r x_{n-1}) \\
&\leq kd(x_n, x_{n-1}) \\
&\leq k^2 d(x_{n-1}, x_{n-2}) \\
&\vdots \\
&\leq k^n d(x_1, x_0)
\end{aligned}$$

So for $n > m$, we have,

$$\begin{aligned}
d(x_n, x_m) &\leq d(x_n, x_{n-1}) + \cdots + d(x_{m+1}, x_m) \\
&\leq (k^{n-1} + \cdots + k^m)d(x_1, x_0) \\
&\leq \frac{k^m}{1-k} d(x_1, x_0)
\end{aligned}$$

Thus we get $\|d(x_n, x_m)\| \leq \frac{k^m}{1-k} K \|d(x_1, x_0)\|$. It follows that $d(x_n, x_m) \to 0$ as $n, m \to \infty$, hence $\{x_n\}$ is a Cauchy sequence. Since $X$ is complete, there exist $x^* \in X$ such that $x_n \to x^*$ as $n \to \infty$. Now observe that with $k := \frac{3Z\beta^r}{1-2Z\beta^r}$,

$$\begin{aligned}
d(T^r x^*, x^*) &\leq d(T^r x_n, T^r x^*) + d(T^r x_n, x^*) \\
&\leq kd(x_n, x^*) + d(x_{n+1}, x^*)
\end{aligned}$$

Thus, $\|d(T^r x^*, x^*)\| \leq K(\|kd(x_n, x^*) + d(x_{n+1}, x^*)\|) \to 0$ as $n \to \infty$. It follows that $T^r x^* = x^*$. Now we show uniqueness of the fixed point. Suppose $y^* \neq x^*$ is such that $y^* = T^r x^*$ for any $r \in \mathbb{N}$. Then since $T$ is a higher-order Hardy-Rogers Type mapping, we deduce that $d(x^*, y^*) \leq 3Z\beta^r d(x^*, y^*)$, but $1 - 3Z\beta^r > 0$, thus, $d(x^*, y^*) = 0$, hence $x^* = y^*$, and uniqueness follows.

### Corollary A.2 1

Let $(X, d)$ be a complete cone metric space, $P$ be a normal cone with normal constant $K$. For $c \in E$ with $0 \ll c$ and $x_0 \in X$ set $B(x_0, c) = \{x \in X : d(x_0, x) \leq c\}$. Suppose that the mapping $T : X \mapsto X$ satisfies the contractive condition $d(T^r x, T^r y) \leq Z\beta^r[d(x, Tx) + d(y, Ty) + d(x, Ty) + d(y, Tx) + d(x, y)]$, for all $x, y \in B(x_0, c)$, where $Z \geq 1$ and $\beta \in [0, \frac{1}{5})$ come from Proposition 1.3[Ampadu, Clement(2016):Higher Order Hardy-Rogers Contraction Mapping Theorem.Unpublished], and $d(T^r x_0, x_0) \leq (1 - \frac{3Z\beta^r}{1-2Z\beta^r})c$. Then $T$ has a unique fixed point in $B(x_0, c)$

## Proof of Corollary A.2 1

By previous theorem we need only prove that $B(x_0, c)$ is complete and $T^r x_0 \in B(x_0, c)$ for all $x \in B(x_0, c)$. Suppose $\{x_n\}$ is Cauchy in $B(x_0, c)$, then it is also Cauchy in $X$. By completeness of $X$, there is $x \in X$ such that $x_n \to x$ as $n \to \infty$. Now notice that

$$d(x_0, x) \leq d(x_n, x_0) + d(x_n, x) \leq d(x_n, x) + c$$

Since, $x_n \to x$ as $n \to \infty$, then taking limits in the above inequality, we conclude that $d(x_0, x) \leq c$. So $B(x_0, c)$ is complete, since $x \in B(x_0, c)$. From the previous theorem, we know that for all $x, y \in X$, $d(T^r x, T^r y) \leq k d(x, y)$, where, $k := \frac{3Z\beta^r}{1-2Z\beta^r}$. Thus, we observe that for every $x \in B(x_0, c)$, $d(x_0, T^r x_0) \leq d(T^r x_0, x_0) + d(T^r x_0, T^r x) \leq (1 - \frac{3Z\beta^r}{1-2Z\beta^r})c + \frac{3Z\beta^r}{1-2Z\beta^r} c = c$. Hence, $T^r x_0 \in B(x_0, c)$.

## Corollary A.3 1

Let $(X, d)$ be a complete cone metric space, $P$ be a normal cone with normal constant $K$. Suppose the mapping $T : X \mapsto X$ satisfies the contractive condition, $d(T^{rm} x, T^{rm} y) \leq Z\beta^r [d(x, Tx) + d(y, Ty) + d(x, Ty) + d(y, Tx) + d(x, y)]$, for all $x, y \in X$, any $r \in \mathbb{N}$, and some positive integer $m$, where $Z \geq 1$ and $\beta \in [0, \frac{1}{5})$ come from Proposition 1.3 [Ampadu, Clement(2016):Higher Order Hardy-Rogers Contraction Mapping Theorem.Unpublished]. Then $T$ has a unique fixed point in $X$.

## Proof of Corollary A.3 1

From Theorem A.1, $T^{rm}$ has a unique fixed point $x^*$. But $T^{rm}(T^r x^*) = T^r(T^{rm} x^*) = T^r x^*$. So $T^r x^*$ is also a fixed point of $T^{rm}$. Hence, $T^r x^* = x^*$, that is, $x^*$ is a fixed point of $T^r$. Since the fixed point of $T^r$ is also a fixed point of $T^{rm}$, the fixed point of $T^r$ is unique.

## Theorem A.4 1

Let $(X, d)$ be a sequentially compact cone metric space, $P$ be a regular cone. Suppose the mapping $T : X \mapsto X$ satisfies the condition, $d(T^r x, T^r y) \leq [d(x, Tx) + d(y, Ty) + d(x, Ty) + d(y, Tx) + d(x, y)]$, for all $x, y \in X$, $x \neq y$. Then $T$ has a unique fixed point in $X$.

### Proof of Theorem A.4 1

Choose $x_0 \in X$. Set $x_1 = T^r x_0$, $x_2 = T^r x_1 = T^{2r} x_0$, $\cdots$, $x_{n+1} = T^r x_n = T^{r(n+1)} x_0$, $\cdots$
If for some $n$, $x_{n+1} = x_n$, then $x_n$ is a fixed point of $T^r$, and we are done. So we assume that for all $n$, $x_{n+1} \neq x_n$. Now notice that

$$d(x_{n+1}, x_{n+2}) < d(x_n, x_{n+1}) + d(x_{n+1}, x_{n+2}) + d(x_n, x_{n+1}) + d(x_n, x_{n+2})$$

From the above, one deduces upon using the triangle inequality, that

$$d(x_{n+1}, x_{n+2}) \leq 2d(x_{n+1}, x_{n+2}) + 3d(x_n, x_{n+1})$$
$$\leq \frac{1}{3} d(x_n, x_{n+1}) + \frac{2}{3} d(x_{n+1}, x_{n+2})$$

It follows that the sequence $\{d_n := d(x_n, x_{n+1})\}$ is decreasing and bounded below by zero. Since $P$ is regular, there is $d^* \in E$ such that $d_n \to d^*$ as $n \to \infty$. From the sequence compactness of $X$, there is a subsequence $\{x_{n_i}\}$ of $\{x_n\}$ and $x^* \in X$ such that $x_{n_i} \to x^*$ as $i \to \infty$. Now observe that for $i = 1, 2, \cdots$, we have,

$$d(x_{n_{i+1}}, T^r x^*) < d(x_{n_i}, x^*) + d(x_{n_i}, T x_{n_i}) + d(x^*, T x^*) + d(x_{n_i}, T x^*) + d(x^*, T x_{n_i})$$
$$= d(x_{n_i}, x^*) + d(x_{n_i}, x_{n_{i+1}}) + d(x^*, T x^*) + d(x_{n_i}, T x^*) + d(x^*, x_{n_{i+1}})$$
$$< 2d(x_{n_i}, x^*) + d(x_{n_i}, x_{n_{i+1}}) + d(x^*, x_{n_{i+1}})$$

From the above, we have that,

$$\|d(x_{n_{i+1}}, T^r x^*)\| < K \|2d(x_{n_i}, x^*) + d(x_{n_i}, x_{n_{i+1}}) + d(x^*, x_{n_{i+1}})\| \to 0$$

Thus it follows that $x^* = \lim_{i \to \infty} x_{n_{i+1}} = T^r x^*$. So $x^*$ is a fixed point of $T^r$. Note by Lemma A.20, that $\lim_{i \to \infty} d(x_{n_{i+1}}, x_{n_i}) = d(T^r x^*, x^*)$, that is, $d^*$ above is given by $d(T^r x^*, x^*)$. As for uniqueness, suppose $y^* \neq x^*$ is another fixed point, then we notice that,

$$d(x^*, y^*) < 3d(x^*, y^*)$$
$$\leq \frac{1}{3} d(T^{2r} x^*, T^{2r} y^*)$$
$$= \frac{1}{3} d(x^*, y^*)$$

Thus we deduce that $\frac{2}{3} d(x^*, y^*) \leq 0$. Consequently, $d(x^*, y^*) = 0$, thus $x^* = y^*$

## 1.4 Exercises

### Generalization of Theorem A.1 1

Taking inspiration from [Rezapour, Sh., Hamlbarani Haghi, R., Some notes on the paper "Cone metric spaces and fixed point theorems of contractive mappings". J. Math. Anal. Appl.345 (2008), no.2, 719-724] show that Theorem A.1 holds if the assumption of normality is dropped. Thus, deduce that Corollary A.2 and Corollary A.3 hold if the assumption of normality is dropped.

### Correction of Theorem 2.8 [J. Math. Anal. Appl.345 (2008), no.2, 719-724] 1

Let $(X, d)$ be a cone metric space. Suppose $T : X \mapsto X$ satisfy $d(Tx, Ty) \leq kd(x, y) + ld(y, Tx)$, where $k, l \geq 0$ and satisfy $k + 2l < 1$. Show $T$ has a unique fixed point, provided $X$ is complete. Deduce that the statement and proof of Theorem 2.8 [Rezapour, Sh., Hamlbarani Haghi, R., Some notes on the paper "Cone metric spaces and fixed point theorems of contractive mappings". J. Math. Anal. Appl.345 (2008), no.2, 719-724] is flawed.

### New Characterization of Contractive Condition in Previous Exercise 1

Let $(X,d)$ be a metric space. A map $T: X \mapsto X$ will be called a $(k, \frac{l}{2})$-contraction if it satisfies $d(Tx, Ty) \leq kd(x,y) + \frac{l}{2}(d(y, Tx) + d(x, Ty))$, where $k, l \geq 0$ and satisfy $k + \frac{l}{2} < 1$. Prove that a $(k, \frac{l}{2})$-contraction mapping has a unique fixed point provided $X$ is complete.

### Alternate Characterization of $(k, \frac{l}{2})$-contraction 1

Let $(X,d)$ be a metric space. A map $T: X \mapsto X$ will be called a $q$-contraction if it satisfies $d(Tx, Ty) \leq q[d(x,y) + d(y, Tx) + d(x, Ty)]$, where $q \geq 0$ and satisfy $q < \frac{1}{2}$. Prove that a $q$-contraction mapping has a unique fixed point provided $X$ is complete.

### Characterization of higher-order $(k, \frac{l}{2})$-contraction 1

Let $(X,d)$ be a metric space and let the self-map $T: X \to X$ satisfy $d(T^r x, T^r y) \leq \sum_{m=0}^{r-1} c_m [d(T^m x, T^{m+1} y) + d(T^m y, T^{m+1} x) + d(T^m x, T^m y)]$ for all $x, y$ in $X$ and $r \in \mathbb{N}$, where $0 \leq c_m < \frac{1}{2}$ for all $0 \leq m \leq r-1$. Prove that $T: X \to X$ is an $r$th order $q$-contraction mapping by mathematical induction on $r \in \mathbb{N}$

### Alternate Characterization of higher-order $(k, \frac{l}{2})$-contraction 1

Let $(X,d)$ be a metric space and $T: X \mapsto X$. Taking inspiration from [Ezearn Fixed Point Theory and Applications (2015) 2015:88], deduce that there exist $W \geq 1$ and $\zeta \in [0, \frac{1}{2})$ such that $d(T^r x, T^r y) \leq W\zeta^r [d(x,y) + d(y, Tx) + d(x, Ty)]$ for any $r \in \mathbb{N}$

### Fixed Point Theorem for higher-order $(k, \frac{l}{2})$-contraction mapping in Cone Metric Space 1

Let $(X,d)$ be a complete cone metric space, $P$ be a normal cone with normal constant $K$. Suppose $T: X \mapsto X$ is the map from the previous exercise, show that $T$ has a unique fixed point in $X$. Deduce further that this result holds if we drop the normality assumption.

## 1.5 References

(1) Huang Long-Guang and Zhang Xian, Cone metric spaces and fixed point theorems of contractive mappings, J. Math. Anal. Appl. 332 (2007) 1468–1476

(2) Ampadu, Clement(2016):Higher Order Hardy-Rogers Contraction Mapping Theorem. Unpublished

(3) Rezapour, Sh., Hamlbarani Haghi, R., Some notes on the paper "Cone metric spaces and fixed point theorems of contractive mappings". J.Math. Anal. Appl.345 (2008), no.2, 719-724

(4) Ezearn Fixed Point Theory and Applications (2015) 2015:88

# Chapter 2

# Best Proximity Point in Regular Cone Metric Spaces for Cyclic Higher-Order Hardy-Rogers Type Contraction

## 2.1 Brief Summary

**Abstract B.1 1**

Taking inspiration from [Eldred, A. A., Veeramani, P., Existence and convergence of best proximity points. J. Math. Anal. Appl. 323 (2006), no.2, 1001-1006; Kirk,W. A., Sirinavasan, P. S., Veeramani, P., Fixed points for mappings satisfying cyclical contractive conditions. Fixed Point Theory 4 (2003), no.1, 79 - 89]. In this chapter, we introduce a notion of cyclic higher-order hardy-rogers-type contraction and obtain some existence of best proximity point for this class of maps

## 2.2 Preliminaries

**Definition B.1 1**

Let $A$ and $B$ be nonempty subsets of a metric space $X$. If there is a pair $(x_0, y_0) \in A \times B$ for which $d(x_0, y_0) = d(A, B)$, where $d(A, B) = \inf\{d(x, y) : x \in A, y \in B\}$, then the pair $(x_0, y_0)$ is called a best proximity pair for $A$ and $B$

**Definition B.2 1**

Let $A$ and $B$ be nonempty subsets of a metric space $X$. Let $T : A \cup B \mapsto A \cup B$ be such that for any $r \in \mathbb{N}$, $T(A) \subseteq_r B$, that is, $T^r(A) \subseteq B$, and $T(B) \subseteq_r A$. We say $x \in A \cup B$ is a $r$-best proximity point for $T$ if $d(x, T^r x) = d(A, B)$

**Remark B.3 1**

If $r = 1$, then the $r$-best proximity point becomes a best proximity point

**Remark B.4 1**

If $r = 1$ and $A = B$, then the $r$-best proximity point becomes a fixed point

# CHAPTER 2. BEST PROXIMITY POINT IN REGULAR CONE METRIC SPACES FOR CYCLIC HIGHER-ORDER HARDY-ROGERS TYPE CONTRACTION

**Definition B.5 1**

Let $A$ and $B$ be nonempty subsets of a metric space $X$. Let $T : A \cup B \mapsto A \cup B$ be such that for any $r \in \mathbb{N}$, $T(A) \subseteq_r B$, that is, $T^r(A) \subseteq B$, and $T(B) \subseteq_r A$. We will say $T : A \cup B \mapsto A \cup B$ is a cyclic higher-order Hardy-Rogers-type contraction if the following hold for all $x \in A$ and $y \in B$ $d(T^r x, T^r y) \leq Z\beta^r [d(x, Tx) + d(y, Ty) + d(x, Ty) + d(y, Tx) + d(x,y)] + (1 - 5Z\beta^r)d(A,B)$ where $Z \geq 1$ and $\beta \in [0, \frac{1}{5})$ come from Proposition 1.3[Ampadu, Clement(2016):Higher Order Hardy-Rogers Contraction Mapping Theorem. Unpublished]

**Notation B.6 1**

$E$ will denote a real Banach space

**Notation B.7 1**

$P$ will denote a subset of $E$

**Definition B.8 1**

[Huang Long-Guang and Zhang Xian, Cone metric spaces and fixed point theorems of contractive mappings, J. Math. Anal. Appl. 332 (2007) 1468–1476] $P$ is called a cone if and only if

(a) $P$ is closed, nonempty, and $P \neq \{0\}$

(b) $a, b \in \mathbb{R}, a, b \geq 0, x, y \in P \Rightarrow ax + by \in P$

(c) $x \in P$ and $-x \in P \Rightarrow x = 0$

**Definition B.9 1**

Given cone $P \subset E$, we define a partial ordering $\leq$ with respect to $P$ by $x \leq y$ iff $y - x \in P$

**Notation B.10 1**

$x < y$ will mean $x \leq y$, but $x \neq y$, while $x \ll y$ will stand for $y - x \in int\, P$, were $int\, P$ denotes the interior of $P$

**Definition B.11 1**

[Huang Long-Guang and Zhang Xian, Cone metric spaces and fixed point theorems of contractive mappings, J. Math. Anal. Appl. 332 (2007) 1468–1476] The cone $P$ is called normal if there is a number $K > 0$ such that for all $x, y \in E$, $0 \leq x \leq y$ implies $\|x\| \leq K\|y\|$. The least positive number satisfying, $0 \leq x \leq y$ implies $\|x\| \leq K\|y\|$, is called the normal constant of $P$

**Definition B.12 1**

[Huang Long-Guang and Zhang Xian, Cone metric spaces and fixed point theorems of contractive mappings, J. Math. Anal. Appl. 332 (2007) 1468–1476] If $\{x_n\}$ is a sequence such that $x_1 \leq x_2 \leq \cdots \leq x_n \leq \cdots \leq y$ for some $y \in E$, and there is $x \in E$ such that $\|x_n - x\| \to 0$ as $n \to \infty$, then the cone $P$ is said to be regular.

Equivalently, the above definition can be stated in the following way

**Definition B.13 1**

[Huang Long-Guang and Zhang Xian, Cone metric spaces and fixed point theorems of contractive mappings, J. Math. Anal. Appl. 332 (2007) 1468–1476] The cone $P$ is regular iff every decreasing sequence which is bounded from below is convergent.

### Lemma B.14 1

[Rezapour, Sh., Hamlbarani Haghi, R., Some notes on the paper "Cone metric spaces and fixed point theorems of contractive mappings". J. Math. Anal. Appl. 345 (2008), no.2, 719-724] Every regular cone is normal.

### Definition B.15 1

[Huang Long-Guang and Zhang Xian, Cone metric spaces and fixed point theorems of contractive mappings, J. Math. Anal. Appl. 332 (2007) 1468–1476] Let $X$ be a nonempty set. Suppose the mapping $d : X \times X \mapsto E$ satisfies

(a) $0 < d(x,y)$ for all $x, y \in X$ and $d(x,y) = 0$ iff $x = y$

(b) $d(x,y) = d(y,x)$ for all $x, y \in X$

(c) $d(x,y) \leq d(x,z) + d(z,y)$ for all $x, y, z \in X$

Then $d$ is called a cone metric on $X$, and $(X,d)$ is called a cone metric space

### Example B.16 1

[Haghi, R. H., Rezapour, Sh., Fixed points of multifunctions on regular cone metric space. Expo. Math. 28 (2010), no.1, 71-77] Let $E = (L^1[0,1], \|\cdot\|_1)$, $P = \{f \in E : f \geq 0\ a.e\}$, $(X, \rho)$ be a metric space, and $d : X \times X \mapsto E$ be defined by $d(x,y) = f_{x,y}$, where $f_{x,y}(t) = \rho(x,y)t^2$. Then $(X,d)$ is a regular cone metric space.

### Definition B.17 1

A map $f : P \mapsto P$ will be called $r$-increasing (strictly $r$-increasing) whenever $x \leq y$ implies $f(x) \leq_r f(y)$, that is, $f^r(x) \leq f^r(y)$ ($x < y$ implies $f(x) <_r f(y)$) for any $r \in \mathbb{N}$

### Definition B.18 1

[Haghi, R. H., Rakocevic, V., Rezapour, Sh., Shahzad, N., Best proximity result in regular cone metric space. Rend. Circ. Mat. Palermo (2) 60 (2011), no.3, 323-327] Let $A$ and $B$ be nonempty subsets of a cone metric space $(X,d)$. An element $p \in P$ is said to be a lower bound for $A \times B$ whenever $p \leq d(a,b)$ for all $(a,b) \in A \times B$. If $p \geq q$ for all lower bound $q$ for $A \times B$, then $p$ is called the greatest lower bound for $A \times B$

### Definition B.19 1

[Haghi, R. H., Rakocevic, V., Rezapour, Sh., Shahzad, N., Best proximity result in regular cone metric space. Rend. Circ. Mat. Palermo (2) 60 (2011), no.3, 323-327] A map $\psi : P \mapsto P$ is called a cone $L$-function whenever $\psi(0) = 0$, $\psi(s) > 0$ for all $s \in P$ with $s \neq 0$ and there exists $\delta_s \gg 0$ such that $\psi(t) \leq s$ for all $s \leq t \leq s + \delta_s$

### Lemma B.20 1

[Haghi, R. H., Rakocevic, V., Rezapour, Sh., Shahzad, N., Best proximity result in regular cone metric space. Rend. Circ. Mat. Palermo (2) 60 (2011), no.3, 323-327] Let $\psi : P \mapsto P$ be a cone $L$-function and $\{s_n\}$ a decreasing sequence in $P$ such that $s_{n+1} < \psi(s_n)$ for all $n \geq 1$. Then $s \to 0$ as $n \to \infty$

## 2.3 Main Results

**Theorem B.1 1**

Let $T : A \cup B \mapsto A \cup B$ be such that for any $r \in \mathbb{N}$, $T(A) \subseteq_r B$, that is, $T^r(A) \subseteq B$, and $T(B) \subseteq_r A$, and $d(T^r x, T^r y) \leq Z\beta^r[d(x,Tx)+d(y,Ty)+d(x,Ty)+d(y,Tx)+d(x,y)]+(1-5Z\beta^r)d(a,b)$ for all $(a,b),(x,y) \in A \times B$, where $Z \geq 1$ and $\beta \in [0, \frac{1}{5})$ come from Proposition 1.3 [Ampadu, Clement(2016):Higher Order Hardy-Rogers Contraction Mapping Theorem. Unpublished]. Then $d(A,B)$ exists.

**Proof of Theorem B.1 1**

Take $x_0 \in A \cup B$ and set $x_{n+1} = T^r x_n$ and $d_{n+1} = d(x_{n+1}, x_n)$ for all $n \geq 1$. Now we notice upon using the triangle inequality that $d_{n+1} \leq Z\beta^r[2d_{n+1} + 3d_n] + (1-5Z\beta^r)d(a,b)$ for all $(a,b) \in A \times B$. From which it follows that $d_{n+1} \leq \frac{3Z\beta^r}{1-2Z\beta^r}d_n + \frac{(1-5Z\beta^r)}{1-2Z\beta^r}d(a,b)$ for all $(a,b) \in A \times B$. It follows that $d_{n+1} \leq d_n$ for all $n \geq 1$. By the regularity of the cone $P$, there exists $p \in P$ such that $d_n \to p$ as $n \to \infty$. Thus, $p \leq d(a,b)$ holds for any $(a,b)$ in $A \times B$. Now if $q$ is a lower bound for $A \times B$ then $q \leq d_n$ for all $n \geq 1$, and so, $q \leq p$. Therefore, $d(A,B) = p$

**Theorem B.2 1**

Suppose that the conditions of the previous theorem hold, $x_0 \in A$ and $x_{n+1} = T^r x_n$ for all $n \geq 1$ and any $r \in \mathbb{N}$. If $\{x_{2n}\}$ has a convergent subsequence in $A$, then there exists $x \in A$ such that $d(x, T^r x) = d(A, B)$.

**Proof of Theorem B.2 1**

Let $\{x_{2n_k}\}$ be a convergent subsequence of $\{x_{2n}\}$ in $A$ with $x_{2n_k} \to x \in A$. Since $p = d(A,B) \leq d(x, x_{2n_k-1}) \leq d(x, x_{2n_k}) + d(x_{2n_k}, x_{2n_k-1})$ for each $k \geq 1$, it follows that $\{d(x_{2n_k}, x_{2n_k-1})\}$ is a subsequence of $d_n$. Hence, $d(x, x_{2n_k-1}) \to p$ as $n \to \infty$. As $p \leq d(T^r x, x_{2n_k}) \leq d(x, x_{2n_k-1})$ for all $k \geq 1$. It follows that $d(x, T^r x) = p = d(A, B)$

Now, we will consider the best proximity points for a pair of mapping $(S,T)$ such that $S,T : A \cup B \mapsto A \cup B$, $S(A) \subseteq_r B$ and $T(B) \subseteq_r A$, for any $r \in \mathbb{N}$

**Theorem B.3 1**

Let $S,T : A \cup B \mapsto A \cup B$, $S(A) \subseteq_r B$ and $T(B) \subseteq_r A$, for any $r \in \mathbb{N}$, and $d(S^r x, T^r y) \leq Z\beta^r[d(x,Sx)+d(y,Ty)+d(x,Ty)+d(y,Sx)+d(x,y)]+(1-5Z\beta^r)d(a,b)$ for all $(a,b),(x,y) \in A \times B$, where $Z \geq 1$ and $\beta \in [0, \frac{1}{5})$ come from Proposition 1.3[Ampadu, Clement(2016):Higher Order Hardy-Rogers Contraction Mapping Theorem. Unpublished]. Then $d(A,B)$ exists.

### Proof of Theorem B.3 1

Take $x_0 \in A$, then $S^r x_0 \in B$ for any $r \in \mathbb{N}$. So there exists $y_0 \in B$ such that $y_0 = S^r x_0$. Now $T^r y_0 \in A$, so there exists $x_1 \in A$ such that $x_1 = T^r x_0$. Inductively, we define sequence $\{x_n\}$ and $\{y_n\}$ in $A$ and $B$, respectively by, $x_{n+1} = T^r y_n$ and $y_n = S^r x_n$. Now set $d_n := d(x_n, S^r x_n)$, and notice that

$$\begin{aligned}d_{n+1} &\leq Z\beta^r[d(x_{n+1}, Sx_{n+1}) + d(y_n, Ty_n) + d(x_{n+1}, Ty_n) + d(y_n, Sx_{n+1}) \\ &\quad + d(x_{n+1}, y_n)] + (1 - 5Z\beta^r)d(a,b) \\ &= Z\beta^r[d_{n+1} + 2d(y_n, x_{n+1}) + d(y_n, y_{n+1})] + (1 - 5Z\beta^r)d(a,b) \\ &\leq Z\beta^r[d_{n+1} + 2d(y_n, x_{n+1}) + d(y_n, x_{n+1}) + d(x_{n+1}, y_{n+1})] + (1 - 5Z\beta^r)d(a,b) \\ &= Z\beta^r[2d_{n+1} + 3d(y_n, x_{n+1})] + (1 - 5Z\beta^r)d(a,b)\end{aligned}$$

From the above one deduces that $d_{n+1} \leq \frac{3Z\beta^r}{1-2Z\beta^r} d(y_n, x_{n+1}) + \frac{1-5Z\beta^r}{1-2Z\beta^r} d(a,b)$.

On the other hand, we notice that,

$$\begin{aligned}d(y_n, x_{n+1}) &\leq Z\beta^r[d(x_n, Sx_n) + d(y_n, Ty_n) + d(x_n, Ty_n) + d(x_n, y_n)] \\ &\quad + (1 - 5Z\beta^r)d(a,b) \\ &= Z\beta^r[3d(x_n, y_n) + 2d(y_n, x_{n+1})] + (1 - 5Z\beta^r)d(a,b)\end{aligned}$$

From the above one deduces that $d(y_n, x_{n+1}) \leq \frac{3Z\beta^r}{1-2Z\beta^r} d_n + \frac{1-5Z\beta^r}{1-2Z\beta^r} d(a,b)$

From the inequalities, $d_{n+1} \leq \frac{3Z\beta^r}{1-2Z\beta^r} d(y_n, x_{n+1}) + \frac{1-5Z\beta^r}{1-2Z\beta^r} d(a,b)$, and, $d(y_n, x_{n+1}) \leq \frac{3Z\beta^r}{1-2Z\beta^r} d_n + \frac{1-5Z\beta^r}{1-2Z\beta^r} d(a,b)$, one concludes that

$$d_{n+1} \leq \frac{(3Z\beta^r)^2}{(1-2Z\beta^r)^2} d_n + \frac{(1-5Z\beta^r)(1+Z\beta^r)}{(1-2Z\beta^r)^2} d(a,b)$$

It follows that $d_{n+1} \leq d_n$ for all $n \geq 1$. By the regularity of the cone $P$, there exists $p \in P$ such that $d_n \to p$ as $n \to \infty$. Thus, $p \leq d(a,b)$ holds for any $(a,b)$ in $A \times B$. Now if $q$ is a lower bound for $A \times B$ then $q \leq d_n$ for all $n \geq 1$, and so, $q \leq p$. Therefore, $d(A,B) = p$

### Theorem B.4 1

Suppose the conditions of the previous theorem hold and the sequence $\{x_n\}$ and $\{y_n\}$ are generated by $x_{n+1} = T^r y_n$ and $y_n = S^r x_n$ for some $x_0 \in A \cup B$. If both $\{x_n\}$ and $\{y_n\}$ have a convergent subsequence in $A$ and $B$ respectively, then there exists $x \in A$ and $y \in B$ such that $d(x, S^r x) = d(A, B) = d(y, T^r y)$

# CHAPTER 2. BEST PROXIMITY POINT IN REGULAR CONE METRIC SPACES FOR CYCLIC HIGHER-ORDER HARDY-ROGERS TYPE CONTRACTION

**Proof of Theorem B.4 1**

Set $d_n = d(x_n, S^r x_n)$. Let $\{y_{n_k}\}$ be a subsequence of $\{y_n\}$ such that $y_{n_k} \to y$. Notice that $p = d(A,B) \leq d(T^r y_{n_k}, y) \leq d(y_{n_k}, y) + d(y_{n_k}, T^r y_{n_k})$ holds for each $k \geq 1$. Since $d(y_{n_k}, T^r y_{n_k}) \leq \frac{3Z\beta^r}{1-2Z\beta^r} d_{n_k} + \frac{(1-5Z\beta^r)(1+Z\beta^r)}{1-2Z\beta^r} d(a,b)$ for all $(a,b) \in A \times B$. It follows that $d(y_{n_k}, T^r y_{n_k}) \leq d_{n_k}$. Since $\{d(S^r x_{n_k}, x_{n_k})\}$ is a subsequence of $\{d_n\}$, it follows that $d(S^r x_{n_k}, x_{n_k}) \to p$ as $k \to \infty$. Thus, $d(y_{n_k}, T^r y_{n_k}) \to p$ as $k \to \infty$. So, $d(T^r y_{n_k}, y) \to p$ as $k \to \infty$. Now for each $k \geq 1$, observe that,

$$\begin{aligned}
d(T^r y, y_{n_k}) &\leq Z\beta^r [d(x_{n_k}, Sx_{n_k}) + d(y, Ty) + d(x_{n_k}, Ty) + d(y, Sx_{n_k}) + d(x_{n_k}, y)] \\
&\quad + (1 - 5Z\beta^r) d(a,b) \\
&= Z\beta^r [d(x_{n_k}, y_{n_k}) + d(y, Ty) + d(x_{n_k}, Ty) + d(y, y_{n_k}) + d(x_{n_k}, y)] \\
&\quad + (1 - 5Z\beta^r) d(a,b) \\
&\leq Z\beta^r [3d(x_{n_k}, y_{n_k}) + 3d(y_{n_k}, y) + 2d(y_{n_k}, Ty)] + (1 - 5Z\beta^r) d(a,b) \\
&\leq Z\beta^r [3d(x_{n_k}, y_{n_k}) + 3d(y_{n_k}, y) + 2d(y_{n_k}, T^r y)] + (1 - 5Z\beta^r) d(a,b)
\end{aligned}$$

From the above inequality one deduces that

$$d(y_{n_k}, T^r y) \leq \frac{3Z\beta^r}{1 - 2Z\beta^r} [d(x_{n_k}, y_{n_k}) + d(y_{n_k}, y)] + \frac{1 - 5Z\beta^r}{1 - 2Z\beta^r} d(a,b)$$

Thus it follows that

$$p = d(A,B) \leq d(y_{n_k}, T^r y) \leq \frac{3Z\beta^r}{1 - 2Z\beta^r} [d(x_{n_k}, y_{n_k}) + d(y_{n_k}, y)] + \frac{1 - 5Z\beta^r}{1 - 2Z\beta^r} d(a,b)$$

for all $(a,b) \in A \times B$. Taking limits in the above inequality one concludes that $d(T^r y, y) = p = d(A,B)$. Similarly, one can show that $d(x, S^r x) = d(A,B)$

Finally the distance of $A$ and $B$ is obtained by considering the pair mapping $(S, T)$ in a regular cone metric space.

**Theorem B.5 1**

Let $\psi : P \mapsto P$ be a cone $L$-function, $S, T : A \cup B \mapsto A \cup B$, $S(A) \subseteq_r B$ and $T(B) \subseteq_r A$, for any $r \in \mathbb{N}$, and $d(S^r x, T^r y) - p < \psi([d(x, Sx) + d(y, Ty) + d(x, Ty) + d(y, Sx) + d(x,y)] - p)$ for all $(x,y) \in A \times B$, with $p < d(x,y)$, where $p$ is a lower bound for $A \times B$. Then $d(A,B) = p$

### Proof of Theorem B.5 1

Let $\{x_n\}$ and $\{y_n\}$ be defined by $x_{n+1} = T^r y_n$ and $y_n = S^r x_n$ for some $(x_0, y_0) \in A \times B$, $n \in \mathbb{N}$, and any $r \in \mathbb{N}$. Also set $d_n := d(x_n, S^r x_n)$. First note that if $\psi$ is the identity, then, we deduce that

$$d(y_n, x_{n+1}) \leq 3d(x_n, y_n) + 2d(y_n, x_{n+1})$$
$$\leq \frac{1}{3} d(x_n, y_n) + \frac{2}{3} d(y_n, x_{n+1})$$

From the above one deduces that, $d(y_n, x_{n+1}) \leq d(x_n, y_n)$. On the other hand we observe that,

$$d_{n+1} - p < \psi([2d_{n+1} + 3d(y_n, x_{n+1})]) - p)$$
$$\leq \psi([\frac{1}{2} d_{n+1} + \frac{1}{3} d(y_n, x_{n+1})]) - p)$$
$$\leq \psi([\frac{1}{2} d_{n+1} + \frac{1}{3} d(y_n, x_n)]) - p)$$
$$\leq \psi([\frac{1}{2} d_{n+1} + d_n] - p)$$
$$\leq \frac{1}{2} d_{n+1} + d_n - p$$

By the regularity of the cone $P$, we have $d_{n+1} \leq d_n$. Hence, there exists $q \in P$ such that $d_n \to p$ as $n \to \infty$. Thus, $p \leq q$. Put $s_n = d_n - p$. Since $s_n > 0$, then $s_{n+1} < \psi[\frac{1}{2} d_{n+1} + s_n] \leq \psi[\frac{3}{2} s_n] \leq \frac{3}{2} s_n$. Thus, $\frac{2}{3} s_{n+1} \leq s_n$; combining with the fact that $\frac{2}{3} s_{n+1} < s_{n+1}$, we conclude that $s_{n+1} \leq s_n$. So by Lemma B.20, we have $s_n \to 0$ as $n \to \infty$, thus, $d_n \to p$ as $n \to \infty$, and so, $d(A, B) = p = q$

## 2.4 Exercises

### Exercise B.1 1

Complete the proof of Theorem B.4 by showing $d(x, S^r x) = d(A, B)$ for any $r \in \mathbb{N}$

### New Characterization of Reich Mapping 1

Deduce from the original definition of a Reich mapping [Canad. Math. Bull. Vol. 14 (1), 1971] that a map $T : X \mapsto X$ satisfying $d(Tx, Ty) \leq \frac{a+b}{2}[d(x, Tx) + d(y, Ty)] + cd(x, y)$ is also a Reich type mapping where $a, b, c \geq 0$ and satisfy $\frac{a+b}{2} + c < 1$.

### Exercise B.3 1

Let $(X, d)$ be a metric space. Prove that if $T : X \mapsto X$ is the map from the previous exercise, then $T$ has a unique fixed point provided $X$ is complete.

### Exercise B.4 1

From the new characterization of the Reich mapping, deduce that a map $T : X \mapsto X$ satisfying $d(Tx, Ty) \leq k[d(x, Tx) + d(y, Ty) + d(x, y)]$, where $k < \frac{1}{2}$, is a Reich type mapping.

The remaining exercises regard characterization of some best proximity point in regular cone metric spaces for various type of cyclic contraction maps. We **do not** consider the map in the sense of [Ezearn Fixed Point Theory and Applications (2015) 2015:88] in the remaining exercises, that is, the map is **not** of the higher-order. We make the following (motivated by the previous exercise)

> **Cyclic Reich Type Contraction 1**
>
> Let $A$ and $B$ be nonempty subsets of a metric space $X$. A map $T : A \cup B \mapsto A \cup B$, $T(A) \subseteq B$, $T(B) \subseteq A$ will be called a cyclic Reich type contraction if for some $k \in (0, \frac{1}{2})$, the condition, $d(Tx, Ty) \leq k[d(x, Tx) + d(y, Ty) + d(x, y)] + (1 - 2k)d(A, B)$, holds for all $x \in A, y \in B$

> **Exercise B.5 1**
>
> Taking inspiration from [Eldred, A. A., Veeramani, P., Existence and convergence of best proximity points. J. Math. Anal. Appl. 323 (2006), no.2, 1001-1006] prove the following: Let $A$ and $B$ be nonempty closed subsets of a complete metric space X. Let $T : A \cup B \mapsto A \cup B$ be a cyclic Reich type contraction. Let $x_0 \in A$ and define $x_{n+1} = Tx_n$. Suppose $\{x_{2n}\}$ has a convergent subsequence in $A$. Then there exists $x \in A$ such that $d(x, Tx) = d(A, B)$

> **Exercise B.6 1**
>
> Prove the following: Let $T : A \cup B \mapsto A \cup B$ be such that $T(A) \subseteq B$, and $T(B) \subseteq A$, and $d(Tx, Ty) \leq \frac{k}{2}[d(x, Tx) + d(y, Ty) + d(x, y)] + (1 - k)d(a, b)$ for all $(a, b), (x, y) \in A \times B$, where $k \in (0, 1)$. Then $d(A, B)$ exists.

> **Exercise B.7 1**
>
> Suppose the conditions of the previous exercise hold, $x_0 \in A$ and $x_{n+1} = Tx_n$ for all $n \geq 1$. If $\{x_{2n}\}$ has a convergent subsequence in $A$, then there exists $x \in A$ such that $d(x, Tx) = d(A, B)$.

> **Exercise B.8 1**
>
> Prove the following: Suppose $S, T : A \cup B \mapsto A \cup B$ be such that $S(A) \subseteq B$, and $T(B) \subseteq A$, and $d(Sx, Ty) \leq \frac{k}{2}[d(x, Sx) + d(y, Ty) + d(x, y)] + (1 - k)d(a, b)$ for all $(a, b), (x, y) \in A \times B$, where $k \in (0, 1)$. Then $d(A, B)$ exists.

> **Exercise B.9 1**
>
> Suppose the conditions of the previous exercise hold and the sequence $\{x_n\}$ and $\{y_n\}$ are generated by $x_{n+1} = Ty_n$ and $y_n = Sx_n$ for some $x_0 \in A \cup B$. If both $\{x_n\}$ and $\{y_n\}$ have a convergent subsequence in $A$ and $B$ respectively, then there exists $x \in A$ and $y \in B$ such that $d(x, Sx) = d(A, B) = d(y, Ty)$

> **Exercise B.10 1**
>
> Let $\psi : P \mapsto P$ be a cone L-function, $S, T : A \cup B \mapsto A \cup B$, $S(A) \subseteq B$ and $T(B) \subseteq A$, and $d(Sx, Ty) - p < \psi([d(x, Sx) + d(y, Ty) + d(x, y)] - p)$ for all $(x, y) \in A \times B$, with $p < d(x, y)$, where $p$ is a lower bound for $A \times B$. Then $d(A, B) = p$

## 2.5 References

(1) Eldred, A. A., Veeramani, P., Existence and convergence of best proximity points. J. Math. Anal. Appl. 323 (2006), no.2, 1001-1006

(2) Kirk,W. A., Sirinavasan, P. S., Veeramani, P., Fixed points for mappings satisfying cyclical contractive conditions. Fixed Point Theory 4 (2003), no.1, 79 - 89

(3) Ampadu, Clement(2016):Higher Order Hardy-Rogers Contraction Mapping Theorem. Unpublished

(4) Huang Long-Guang and Zhang Xian, Cone metric spaces and fixed point theorems of contractive mappings, J. Math. Anal. Appl. 332 (2007)1468–1476

(5) Rezapour, Sh., Hamlbarani Haghi, R., Some notes on the paper "Cone metric spaces and fixed point theorems of contractive mappings". J. Math. Anal. Appl. 345 (2008), no.2, 719-724

(6) Haghi, R. H., Rezapour, Sh., Fixed points of multifunctions on regular cone metric space. Expo. Math. 28 (2010), no.1, 71-77

(7) Haghi, R. H., Rakocevic, V., Rezapour, Sh., Shahzad, N., Best proximity result in regular cone metric space. Rend. Circ. Mat. Palermo (2) 60 (2011), no.3, 323-327

(8) Canad. Math. Bull. Vol. 14 (1), 1971

(9) Ezearn Fixed Point Theory and Applications (2015) 2015:88

# Chapter 3

# Fixed Point Theorems for Ordered g-Weak Higher-Order Hardy-Rogers Type Contraction in Ordered Cone Rectangular Metric Spaces

## 3.1 Brief Summary

> **Abstract C.1 1**
>
> In this chapter we introduce a notion of $r$-$g$-weak higher-order Hardy-Rogers type contraction and prove some common $r$-fixed point theorems in ordered cone rectangular metric spaces.

## 3.2 Preliminaries

> **Notation C.1 1**
>
> $E$ will denote a real Banach space

> **Notation C.2 1**
>
> $P$ will denote a subset of $E$

> **Definition C.3 1**
>
> [Huang Long-Guang and Zhang Xian, Cone metric spaces and fixed point theorems of contractive mappings, J. Math. Anal. Appl. 332 (2007) 1468–1476] $P$ is called a cone if and only if
>
> (a) $P$ is closed, nonempty, and $P \neq \{0\}$
>
> (b) $a, b \in \mathbb{R}$, $a, b \geq 0$, $x, y \in P \Rightarrow ax + by \in P$
>
> (c) $x \in P$ and $-x \in P \Rightarrow x = 0$

> **Definition C.4 1**
>
> Given cone $P \subset E$, we define a partial ordering $\leq$ with respect to $P$ by $x \leq y$ iff $y - x \in P$

## Notation C.5 1

$x < y$ will mean $x \leq y$, but $x \neq y$, while $x \ll y$ will stand for $y - x \in int\, P$, were $int\, P$ denotes the interior of $P$

## Definition C.6 1

[Huang Long-Guang and Zhang Xian, Cone metric spaces and fixed point theorems of contractive mappings, J. Math. Anal. Appl. 332 (2007) 1468–1476] The cone $P$ is called normal if there is a number $K > 0$ such that for all $x, y \in E$, $0 \leq x \leq y$ implies $\|x\| \leq K\|y\|$. The least positive number satisfying, $0 \leq x \leq y$ implies $\|x\| \leq K\|y\|$, is called the normal constant of $P$

## Definition C.7 1

[Huang Long-Guang and Zhang Xian, Cone metric spaces and fixed point theorems of contractive mappings, J. Math. Anal. Appl. 332 (2007) 1468–1476] Let $X$ be a nonempty set. Suppose the mapping $d : X \times X \mapsto E$ satisfies

(a) $0 < d(x, y)$ for all $x, y \in X$ and $d(x, y) = 0$ iff $x = y$

(b) $d(x, y) = d(y, x)$ for all $x, y \in X$

(c) $d(x, y) \leq d(x, z) + d(z, y)$ for all $x, y, z \in X$

Then $d$ is called a cone metric on $X$, and $(X, d)$ is called a cone metric space

In the sequel we will need the following

## Remark C.8 1

[G. Jungck, S. Radenovic, S. Radojevic, and V. Rakocevıc,"Common fixed point theorems for weakly compatible pairs on cone metric spaces," Fixed Point Theory and Applications, vol. 2009, Article ID 643840, pp. 1–13, 2009] Let $P \subseteq E$ be a cone, and $a, b, c \in P$, then we have the following

(a) If $a \leq b$ and $b \ll c$, then $a \ll c$

(b) If $a \ll b$ and $b \ll c$, then $a \ll c$

(c) If $0 \leq u \ll c$ for each $c \in int\, P$, then $u = 0$

(d) If $c \in int\, P$ and $a_n \to 0$, then there exists $n_0 \in \mathbb{N}$ such that, for all $n > n_0$, we have $a_n \ll c$

(e) If $0 \leq a_n \leq b_n$ for each $n$ and $a_n \to a$, $b_n \to b$, then $a \leq b$

(f) If $a \leq \lambda a$, where $0 \leq \lambda < 1$, then $a = 0$

### Definition C.9 1

[A. Azam, M. Arshad, and I. Beg, "Banach contraction principle on cone rectangular metric spaces," Applicable Analysis and Discrete Mathematics, vol. 3, no. 2, pp. 236–241, 2009] Let $X$ be a nonempty set. A mapping $d : X \times X \mapsto E$ is a called a cone rectangular metric on $X$ if it satisfies the following

(a) $0 \leq d(x,y)$ for all $x,y \in X$ and $d(x,y) = 0$ iff $x = y$

(b) $d(x,y) = d(y,x)$ for all $x,y \in X$

(c) $d(x,y) \leq d(x,z) + d(z,y)$ for all $x,y,z \in X$

(d) $d(x,y) \leq d(x,w) + d(w,z) + d(z,y)$ for all $x,y \in X$ and for all distinct points $w,z \in X\text{-}\{x,y\}$

We say $(X,d)$ is a cone rectangular metric space

### Definition C.10 1

Let $\{x_n\}$ be a sequence in $(X,d)$ and $x \in X$. If for every $c \in E$, with $0 \ll c$ there is a natural number $n_0$ such that for all $n > n_0$, $d(x_n, x) \ll c$, then $\{x_n\}$ is said to be convergent to $x \in X$

### Definition C.11 1

If for every $c \in E$ with $0 \ll c$ there is a natural number $n_0$ such that for all $n > n_0$ and a natural number $m$, we have $d(x_n, x_m) \ll c$, then $\{x_n\}$ is called a Cauchy sequence in $(X,d)$

### Definition C.12 1

If every Cauchy sequence is convergent in $(X,d)$, then $(X,d)$ is called a complete cone rectangular metric space

### Remark C.13 1

If the underlying cone is normal then $(X,d)$ is called a normal cone rectangular metric space

### Remark C.14 1

For examples of a cone rectangular metric and a non-normal cone rectangular metric the reader should see [S.K. Malhotra et.al, The Scientific World Journal, Volume 2013, Article ID 810732, 7 pages, http://dx.doi.org/10.1155/2013/810732]

### Definition C.15 1

Let $f$ and $g$ be self-mappings of a nonempty set $X$ and $C_r(f,g) = \{x \in X : f^r x = g^r x\}$ for any $r \in \mathbb{N}$. The pair $(f,g)$ will be called $r$-weakly compatible if $fg^r x = gf^r x$ for all $x \in C_r(f,g)$ and any $r \in \mathbb{N}$. If $w = f^r x = g^r x$ for some $x \in X$, then $x$ will be called a $r$-coincidence point of $f$ and $g$, and $w$ will be called a $r$-point of coincidence of $f$ and $g$

### Definition C.16 1

If a nonempty set $X$ is equipped with partial order $"\sqsubseteq"$ and $d : X \times X \mapsto E$ is a cone rectangular metric, then, $(X, \sqsubseteq, d)$ is called an ordered cone rectangular metric space

### Definition C.17 1

Let $f,g : X \mapsto X$ be two mappings. The mapping $f$ will be called $r$-nondecreasing with respect to "$\sqsubseteq$" if for each $x,y \in X$, $x \sqsubseteq y$ implies $fx \sqsubseteq_r fy$, that is, $f^r x \sqsubseteq f^r y$

### Definition C.18 1

The mapping $f$ will be called $r$-$g$-nondecreasing if for each $x,y \in X$, $gx \sqsubseteq_r gy$ implies $fx \sqsubseteq_r fy$

### Definition C.19 1

$A \subseteq X$ is called well-ordered if all the elements of $A$ are comparable, that is, for all $x,y \in A$, either $x \sqsubseteq y$ or $y \sqsubseteq x$

### Definition C.20 1

$A \subseteq X$ will be called $r$-$g$-well ordered if all the elements of $A$ are $r$-$g$-comparable, that is, for all $x,y \in A$, either $gx \sqsubseteq_r gy$ or $gy \sqsubseteq_r gx$

### Remark C.21 1

If $r = 1$, then $r$-$g$-well orderedness reduces to $g$-well orderedness

### Remark C.22 1

If $g$ is the identity, then $r$-$g$-well orderedness reduces to well-orderedness

### Example C.23 1

Let $X = \{0,1,2,3,4\}$, let "$\sqsubseteq$" be a partial order relation on $X$ defined by $\sqsubseteq = \{(0,0),(1,1),(2,2),(3,3),(4,4),(1,2),(2,3),(1,3),(1,4)\}$. Let $A = \{0,1,3\}$ and $B = \{1,4\}$, and $g : X \mapsto X$ be defined by, $g^r(0) = 1$ if $r = 1$; $g^r(0) = 2$ if $r = 2$; $g^r(0) = 3$ if $r \in \mathbb{N} - \{1,2\}$; $g^r(1) = 2$ if $r = 1$; $g^r(1) = 3$ if $r \in \mathbb{N} - \{1\}$; $g^r(2) = 3$ for any $r \in \mathbb{N}$; $g^r(3) = 3$ for any $r \in \mathbb{N}$; $g^r(4) = 0$ for $r = 1$; $g^r(4) = 1$ for $r = 2$; $g^r(4) = 2$ for $r = 3$; $g^r(4) = 3$ for any $r \in \mathbb{N} - \{1,2,3\}$. Consequently, the following is clear

(a) $A$ is not well ordered

(b) $B$ is well ordered

(c) $A$ is $r$-$g$-well ordered for any $r \in \mathbb{N}$

(d) $B$ is $r$-$g$- well ordered if and only if $r \in \mathbb{N} - \{1,2,3\}$

By comparing with Example 9 [S.K Malhotra et.al, The Scientific World Journal, Volume 2013, Article ID 810732, 7 pages, http://dx.doi.org/10.1155/2013/810732] it follows that the notion of $r$-$g$-well orderedness is more general than the notion of $g$-well orderedness

### Definition C.24 1

Let $(X, \sqsubseteq, d)$ be an ordered cone rectangular metric space and $f,g : X \mapsto X$ be two mappings. The map $f$ will be called an ordered $r$-$g$-weak higher-order Hardy-Rogers type contraction if $d(f^r x, f^r y) \leq Z\beta^r[d(g^r x, fx) + d(g^r y, fy) + d(g^r x, fy) + d(g^r y, fx) + d(g^r x, g^r y)]$, for all $x,y \in X$ with $gx \sqsubseteq_r gy$, where $Z \geq 1$ and $\beta \in [0, \frac{1}{5})$ come from Proposition 1.3[Ampadu, Clement(2016):Higher Order Hardy-Rogers Contraction Mapping Theorem. Unpublished]

> **Remark C.25 1**
>
> If $g$ is the identity in the previous definition, then ordered $r$-$g$-weak higher-order Hardy-Rogers type contraction reduces to ordered higher-order Hardy-Rogers type contraction

> **Remark C.26 1**
>
> If in addition to the previous remark, the inequality in the previous definition holds only for all $x, y \in X$ then ordered $r$-$g$-weak higher-order Hardy-Rogers type contraction reduces to higher-order Hardy-Rogers type contraction

> **Remark C.27 1**
>
> If the inequality in the previous definition holds only for all $x, y \in X$ then ordered $r$-$g$-weak higher-order Hardy-Rogers type contraction reduces to $r$-$g$-weak higher-order Hardy-Rogers type contraction

## 3.3 Main Results

> **Theorem C.1 1**
>
> Let $(X, \sqsubseteq, d)$ be an ordered cone rectangular metric space, $f, g : X \mapsto X$ be two mappings such that $f(X) \subseteq_r g(X)$, that is, $f^r(X) \subseteq g^r(X)$ for any $r \in \mathbb{N}$ and $g(X)$ is $r$-complete, that is, $g^r(X)$ is complete for any $r \in \mathbb{N}$. Suppose the following conditions are satisfied
>
> (a) $f$ is an ordered $r$-$g$-weak higher-order Hardy-Rogers type contraction
>
> (b) $f$ is $r$-$g$-nondecreasing
>
> (c) there exists $x_0 \in X$ such that $gx_0 \sqsubseteq_r fx_0$
>
> (d) if $\{g^r x_n\}$ is a nondecreasing sequence in $X$ converging to some $g^r z$ for any $r \in \mathbb{N}$, then $gx_n \sqsubseteq_r gz$ for all $n \in \mathbb{N}$ and $gz \sqsubseteq_r ggz$
>
> Then, $f$ and $g$ have a $r$-coincidence point. Furthermore if $f$ and $g$ are $r$-weakly compatible, then they have a common $r$-fixed point. In addition, the set of common $r$-fixed points of $f$ and $g$ is $r$-$g$-well ordered if and only if the common $r$-fixed points of $f$ and $g$ is unique.

> **Proof of Theorem C.1 1**
>
> Let $x_0 \in X$ and define a sequence $\{y_n\}$ as follows: let $f^r x_0 = g^r x_1 = y_1$. As $gx_0 \sqsubseteq_r fx_0$, we have $gx_0 \sqsubseteq_r gx_1$, and as $f$ is $r$-$g$-nondecreasing, we obtain $fx_0 \sqsubseteq_r fx_1$. Now let $f^r x_1 = g^r x_2 = y_2$. Since $gx_1 \sqsubseteq_r gx_2$ and $f$ is $r$-$g$-nondecreasing, we obtain $gx_1 \sqsubseteq_r gx_2$. On repeating this process we obtain for every $n \in \mathbb{N}$,
>
> $$fx_0 \sqsubseteq_r fx_1 \sqsubseteq_r \cdots \sqsubseteq_r fx_n \sqsubseteq_r fx_{n+1} \sqsubseteq_r \cdots$$
> $$gx_0 \sqsubseteq_r gx_1 \sqsubseteq_r \cdots \sqsubseteq_r gx_n \sqsubseteq_r gx_{n+1} \sqsubseteq_r \cdots$$
>
> Thus, $\{y_n\} = \{g^r x_n\} = \{f^r x_{n-1}\}$ is a nondecreasing sequence with respect to $\sqsubseteq$. Now we show that $f$ and $g$ have a $r$-point of coincidence. If $y_n = y_{n+1}$ for any $n \in \mathbb{N}$, then $y_n = g^r x_n = f^r x_n$, therefore $y_n$ is a $r$-point of coincidence of $f$ and $g$ with $r$-coincidence point $x_n$. Therefore, we assume that $y_n \neq y_{n+1}$ for all $n \in \mathbb{N}$. As $gx_n \sqsubseteq_r gx_{n+1}$ for all $n \in \mathbb{N}$ and $f$ is an ordered $r$-$g$-weak higher-order Hardy-Rogers type contraction, one deduces upon setting $d_n = d(y_n, y_{n+1})$, that $d_n \leq \frac{3Z\beta^r}{1-2Z\beta^r} d_{n-1}$. Put $\zeta := \frac{3Z\beta^r}{1-2Z\beta^r}$. By repeating this process we obtain that $d_n \leq \zeta^n d_0$ for all $n \in \mathbb{N}$. If $y_n = y_{n+p}$ for any $n \in \mathbb{N}$ and positive integer $p > 1$, then as $y_n \sqsubseteq y_{n+2}$ and $f$ is an ordered $r$-$g$-weak higher-order Hardy-Rogers type contraction, one deduces upon setting $d_n = d(y_n, y_{n+1})$, that $d_n \leq \zeta d_{n+p-1}$. Repeating this process $p$ times, we obtain $d_n \leq \zeta^p d_n < d_n$, a contradiction. Therefore, we assume that $y_n \neq y_m$ for all distinct $n, m \in \mathbb{N}$. Again as $y_n \sqsubseteq y_{n+2}$, then using the facts that $f$ is an ordered $r$-$g$-weak higher-order Hardy-Rogers type contraction and $d_n \leq \zeta^n d_0$ for all $n \in \mathbb{N}$, we deduce, upon setting $d_n = d(y_n, y_{n+1})$, that
>
> $$\begin{aligned} d(y_n, y_{n+2}) &\leq \zeta d_{n-1} + \zeta d_{n+1} \\ &\leq \zeta^n d_0 + \zeta^{n+2} d_0 \\ &= (1 + \zeta^2) \zeta^n d_0 \end{aligned}$$
>
> Let $\beta = 1 + \zeta^2 \geq 0$, then $d(y_n, y_{n+2}) \leq \beta \zeta^n d_0$. Now for the sequence $\{y_n\}$, we consider $d(y_n, y_{n+p})$ in two cases. If $p$ is odd say $2m+1$, then we deduce that $d(y_n, y_{n+2m+1}) \leq \frac{\zeta^n}{1-\zeta} d_0$. If $p$ is even, say $2m$, then we deduce that $d(y_n, y_{n+2m}) \leq \frac{\zeta^n}{1-\zeta} d_0 + \beta \zeta^n d_0$. As $\beta \geq 0$ and $0 \leq \zeta < 1$, we have, $\frac{\zeta^n}{1-\zeta} d_0 \to 0$ and $\beta \zeta^n d_0 \to 0$. Thus, it follows that for every $c \in E$ with $0 \ll c$ there exists a natural number $n_0$ such that $d(y_n, y_{n+2m+1}) \ll c$ and $d(y_n, y_{n+2m}) \ll c$ for all $n > n_0$. Thus $\{y_n\} = \{g^r x_n\}$ is a Cauchy sequence in $g^r(X)$. As $g^r(X)$ is complete, there exists $z, u \in X$ such that $y_n, g^r x_n, f^r x_{n-1}, g^r z \to u$ as $n \to \infty$. We show $f^r z = g^r z = u$. First notice that
>
> $$\begin{aligned} d(u, f^r z) &\leq d(u, y_n) + d(y_n, y_{n+1}) + d(y_{n+1}, f^r z) \\ &= d(u, y_n) + d_n + d(y_{n+1}, f^r z) \end{aligned}$$
>
> From (d) of the theorem we have $gx_n \sqsubseteq_r gz$, that is, $y_n \sqsubseteq u$, therefore, since $f$ is an ordered $r$-$g$-weak higher-order Hardy-Rogers type contraction, we deduce the following $d(y_{n+1}, f^r z) \leq \frac{3Z\beta^r}{1-2Z\beta^r} d(y_n, u) + \frac{3Z\beta^r}{1-2Z\beta^r} d_n$. It follows that for every $c \in E$ with $0 \ll c$ there is a natural number $n_1$ such that $d(y_n, u) \ll \frac{c(1-2Z\beta^r)}{6Z\beta^r}$ and $d_n \ll \frac{c(1-2Z\beta^r)}{6Z\beta^r}$ for all $n > n_1$. Therefore we deduce that $d(y_{n+1}, fz) \ll c$ for all $n > n_1$ and every $c \in E$ with $0 \ll c$. Consequently, $d(u, f^r z) = 0$, that is, $f^r z = g^r z = u$. Thus, $z$ is a $r$-coincidence point and $u$ is a $r$-point of coincidence of $f$ and $g$. Now suppose that $f$ and $g$ are $r$-weakly compatible, then we have $f^r u = f^r(g^r z) = g^r(f^r z) = g^r u$. As $gz \sqsubseteq_r ggz$ and $f$ is an ordered $r$-$g$-weak higher-order Hardy-Rogers type contraction, we deduce that

### Proof of Theorem C.1 Continued 1

$$d(u, f^r u) \leq 3Z\beta^r d(g^r z, f^r gz)$$
$$= 3Z\beta^r d(u, f^r u)$$

However, $1 - 3Z\beta^r > 0$, thus, $d(u, f^r u) = 0$, hence $u = f^r u = g^r u$. So $u$ is common $r$-fixed point of $f$ and $g$. Finally, suppose the set of common $r$-fixed points of $f$ and $g$ is $r$-$g$-well-ordered and $u, v$ are elements of this set. As the set of common $r$-fixed points $r$-$g$-well-ordered, let, for example, $gu \sqsubseteq_r gv$, then since $f$ is an ordered $r$-$g$-weak higher-order Hardy-Rogers type contraction, one deduces that $d(u, v) \leq 3Z\beta^r d(u, v)$, but $1 - 3Z\beta^r > 0$, thus, $d(u, v) = 0$, that is, $u = v$. It follows that the common $r$-fixed point of $f$ and $g$ is unique. For the converse, let the common $r$-fixed point be unique, then the set of common $r$-fixed point of $f$ and $g$ will be a singleton and therefore $r$-$g$-well-ordered

If $g$ is the identity in the previous theorem, then we obtain the following

### Corollary C.2 1

Let $(X, \sqsubseteq, d)$ be an ordered cone rectangular metric space, $f : X \mapsto X$ be a mapping such that the following conditions are satisfied

(a) $f$ is an ordered higher-order Hardy-Rogers type contraction

(b) $f$ is $r$-nondecreasing with respect to "$\sqsubseteq$"

(c) there exists $x_0 \in X$ such that $x_0 \sqsubseteq f^r x_0$

(d) if $\{x_n\}$ is a nondecreasing sequence in $X$ converging to some $z$, then $x_n \sqsubseteq z$ for all $n$

Then, $f$ has a $r$-fixed point. In addition, the set of $r$-fixed points of $f$ is well-ordered if and only if the $r$-fixed point of $f$ is unique.

### Example C.3 1

Let $X = \{1, 2, 3, 4\}$ and $E = C^1_\mathbb{R}[0, 1]$ with $\|x\| = \|x\|_\infty + \|x'\|_\infty$, $P = \{x(t) : x(t) \geq 0 \text{ for } t \in [0, 1]\}$. Define $d : X \times X \mapsto E$ as follows: $d(1, 2) = d(2, 1) = 3e^t$, $d(2, 3) = d(3, 2) = d(1, 3) = d(3, 1) = e^t$, $d(1, 4) = d(4, 1) = d(2, 4) = d(4, 2) = d(3, 4) = d(4, 3) = 4e^t$, $d(x, y) = 0$ if $x = y$. Then $(X, d)$ is a complete nonnormal cone rectangular metric space but not a cone metric space. Define mappings $f, g : X \mapsto X$ and partial order $\sqsubseteq$ on $X$ as follows: $f^r(1) = 1$, $f^r(2) = 3$, $f^r(3) = 3$, $f^r(4) = 1$, $g^r(1) = 1$, $g^r(2) = 2$, $g^r(3) = 2$, $g^r(4) = 3$ if $r = 1$, $g^r(4) = 2$, if $r \in \mathbb{N} - \{1\}$, $\sqsubseteq = \{(1,1), (2,2), (3,3), (4,4), (1,2), (1,3)\}$. We only need to check that $f$ is a (modified) ordered $r$-$g$-weak higher-order Hardy-Rogers type contraction for $(x, y) = (1, 2), (1, 4)$. Now, $d(f^r(1), f^r(2)) = d(1, 3) = e^t$. On the other hand, $d(g^r(1), g^r(2)) + d(g^r(1), f^r(1)) + d(g^r(2), f^r(2)) = d(1, 2) + d(1, 1) + d(2, 3) = 3e^t + e^t = 4e^t$. Thus, we observe that $e^t = d(f^r(1), f^r(2)) \leq Q\gamma^r[d(g^r(1), g^r(2)) + d(g^r(1), f^r(1)) + d(g^r(2), f^r(2))] = 4e^t Q\gamma^r$. Thus, if $Q \geq 1$ is modified from Proposition 1.3[Ampadu, Clement(2016):Higher Order Hardy-Rogers Contraction Mapping Theorem. Unpublished], then we have equality iff $Q = 1$ and $\gamma^r = \frac{1}{4}$. Thus, there exist $Q \geq 1$ and $\gamma \in [\frac{1}{4}, 1)$ such that $f$ is a (modified) ordered $r$-$g$-weak higher-order Hardy-Rogers type contraction. On the other hand, $d(f^r(1), f^r(4)) = d(1, 1) = 0$, thus given any modification from Proposition 1.3[Ampadu, Clement(2016):Higher Order Hardy-Rogers Contraction Mapping Theorem. Unpublished], $f$ will be a (modified) ordered $r$-$g$-weak higher-order Hardy-Rogers type contraction. Moreover all the conditions of the previous theorem are satisfied and 1 is the unique common $r$-fixed point of $f$ and $g$

### Example C.4 1

Let $(X, d)$ be a cone rectangular metric space as in the previous example. Then $(X, d)$ is a complete nonnormal cone rectangular metric space but not a cone metric space. Define mappings $f, g : X \mapsto X$ and partial order "$\sqsubseteq$" on $X$ as follows: $\sqsubseteq = \{(1,1), (2,2), (3,3), (4,4), (2,4), (2,3), (1,3)\}$, $f^r(1) = 1$, $f^r(2) = 2$, $f^r(3) = 2$, $f^r(4) = 1$ if $r = 1$, $f^r(4) = 2$ if $r \in \mathbb{N} - \{1\}$, $g^r(1) = 1$, $g^r(2) = 3$ if $r$ is odd, $g^r(2) = 2$ if $r$ is even, $g^r(3) = 2$ if $r$ is odd, $g^r(3) = 3$ if $r$ is even, $g^r(4) = 4$. We only need to check that $f$ is a (modified) ordered $r$-$g$-weak higher-order Hardy-Rogers type contraction for $(x, y) = (3, 4), (3, 2), (1, 2)$. Now for $r \in \mathbb{N} - \{1\}$, $d(f^r(3), f^r(4)) = d(2, 2) = 0$, thus, given any modification from Proposition 1.3[Ampadu, Clement(2016):Higher Order Hardy-Rogers Contraction Mapping Theorem. Unpublished], $f$ will be a (modified) ordered $r$-$g$-weak higher-order Hardy-Rogers type contraction. Similarly, $d(f^r(3), f^r(2)) = d(2, 2) = 0$ and $d(f^r(1), f^r(2)) = d(2, 2) = 0$, thus, given any modification from Proposition 1.3[Ampadu, Clement(2016):Higher Order Hardy-Rogers Contraction Mapping Theorem. Unpublished], $f$ will be a (modified) ordered $r$-$g$-weak higher-order Hardy-Rogers type contraction. If $r$ is odd, then 3 is a $r$-coincidence point of of $f$ and $g$, but $f^r g^r(3) \neq g^r f^r(3)$. However if $r$ is even then $2 = f^r(3) \neq g^r(3) = 3$, thus 3 is not a $r$-coincidence point of $f$ and $g$, but $f^r g^r(3) = g^r f^r(3)$. This shows the crucial role of $r$-weak compatibility of mappings for the existence of common $r$-fixed point in the previous theorem.

### Theorem C.5 1

Let $(X, \sqsubseteq, d)$ be an ordered cone rectangular metric space and $f, g : X \mapsto X$ be two mappings such that $f(X) \subseteq_r g(X)$. Suppose that the following conditions are satisfied

(a) $f$ is an ordered $r$-$g$-weak higher-order Hardy-Rogers contraction

(b) there exists $u \in X$ such that $gu \sqsubseteq_r fu$ and $d(g^r u, f^r u) \leq d(g^r x, f^r x)$ for all $x \in X$

Then, $f$ and $g$ have a $r$-coincidence point. Furthermore, if $f$ and $g$ are $r$-weakly compatible, then they have a common $r$-fixed point. In addition the set of common $r$-fixed points of $f$ and $g$ is $r$-$g$-well ordered if and only if the common $r$-fixed point of $f$ and $g$ is unique.

### Proof of Theorem C.5 1

Let $F(x) = d(g^r x, f^r x)$ for all $x \in X$ and $g^r z = f^r u$ for some $z \in X$, then $F(u) \leq F(z)$ for all $x \in X$. If $F(u) = 0$, then $g^r u = f^r u$, that is, $u$ is a $r$-coincidence point of $f$ and $g$. If $0 < F(u)$, then by assumption (b), $gu \sqsubseteq_r fu$, so $gu \sqsubseteq_r gz$, and by (a), we obtain $F(z) = d(g^r z, f^r z) = d(f^r u, f^r z)$. Since $f$ is an ordered $r$-$g$-weak higher-order Hardy-Rogers type contraction, we deduce that $F(z) \leq \frac{3Z\beta^r}{1-2Z\beta^r} F(u) < F(u)$, a contradiction. Thus, $F(u) = 0$, that is, $g^r u = f^r u$, and so $u$ is a $r$-coincidence point of $f$ and $g$. The existence, necessary and sufficient condition for uniqueness of common $r$-fixed point follows from a similar process as used in Theorem C.1

## 3.4 Exercises

### Exercise C.1 1

Taking inspiration from [S.K. Malhotra et.al, SOME FIXED POINT THEOREMS FOR ORDERED REICH TYPE CONTRACTIONS IN CONE RECTANGULAR METRIC SPACES, Acta Math. Univ. Comenianae Vol. LXXXII, 2 (2013), pp. 165-175] prove Corollary C.2 in its entirety

# CHAPTER 3. FIXED POINT THEOREMS FOR ORDERED G-WEAK HIGHER-ORDER HARDY-ROGERS TYPE CONTRACTION IN ORDERED CONE RECTANGULAR METRIC SPACES

> **Exercise C.2 1**
>
> Taking inspiration from [S.K. Malhotra et.al, SOME FIXED POINT THEOREMS FOR ORDERED REICH TYPE CONTRACTIONS IN CONE RECTANGULAR METRIC SPACES, Acta Math. Univ. Comenianae Vol. LXXXII, 2 (2013), pp. 165-175] prove Theorem C.5 holds, if $g$ is the identity

> **Exercise C.3 1**
>
> Let $X$, $E$, $\|x\|$, $P$, "$\sqsubseteq$", and $d : X \times X \mapsto X$ be defined as in Example 3[S.K. Malhotra et.al, SOME FIXED POINT THEOREMS FOR ORDERED REICH TYPE CONTRACTIONS IN CONE RECTANGULAR METRIC SPACES, Acta Math. Univ. Comenianae Vol. LXXXII, 2 (2013), pp. 165-175]. Define $f : X \mapsto X$ for any $r \in \mathbb{N}$ as $f^r(1) = 1$; $f^r(2) = 2$; $f^r(3) = 4$ if $r = 1$; $f^r(3) = 1$ if $r \in \mathbb{N} - \{1\}$; $f^r(4) = 2$ if $r = 1$; $f^r(4) = 1$ if $r \in \mathbb{N} - \{1\}$. Deduce that $f$ is a (modified) ordered higher-order Hardy-Rogers type contraction

The remaining exercises regard obtaining some fixed point and common fixed point theorems in ordered cone b-metric spaces inspired by the higher-order Hardy Rogers type map

> **Exercise C.4 1**
>
> Deduce that in a b-metric space with coefficient $s \geq 1$, a map $T : X \mapsto X$ satisfying the contractive condition, $d(T^r x, T^r y) \leq Z\beta^r[d(x, Tx) + d(y, Ty) + d(x, Ty) + d(y, Tx) + d(x, y)]$, for all $x, y \in X$, is a higher-order Hardy-Rogers type map if and only if there exists $Z \geq 1$ and $\beta \in [0, \frac{6}{5(s^2+3s+2)})$ modified from Proposition 1.3 [Ampadu, Clement(2016):Higher Order Hardy-Rogers Contraction Mapping Theorem. Unpublished]

> **Exercise C.5 1**
>
> Taking inspiration from [Abusalim et.al, Fixed Point and Common Fixed Point Theorems on Ordered Cone b-Metric Spaces, Abstract and Applied Analysis, Volume 2013, Article ID 815289, 7 pages, http://dx.doi.org/10.1155/2013/815289] prove the following: Let $(X, \sqsubseteq)$ be a partially ordered set and suppose there exists a cone b-metric $d$ in $X$ such that $(X, d)$ is a complete cone b-metric space with coefficient $s \geq 1$ relative to a solid cone $P$. Let $T : X \mapsto X$ be a continuous and nondecreasing mapping with respect to "$\sqsubseteq$". Suppose that the following hold
>
> (a) there exist $Z \geq 1$ and $\beta \in [0, \frac{6}{5(s^2+3s+2)})$ modified from Proposition 1.3 [Ampadu, Clement(2016):Higher Order Hardy-Rogers Contraction Mapping Theorem
>
> (b) $d(T^r x, T^r y) \leq Z\beta^r[d(x, Tx) + d(y, Ty) + d(x, Ty) + d(y, Tx) + d(x, y)]$, for all $x, y \in X$ with $x \sqsubseteq y$
>
> (c) there exists $x_0 \in X$ such that $x_0 \sqsubseteq_r Tx_0$, that is, $x_0 \sqsubseteq T^r x_0$
>
> Then $T$ has a $r$-fixed point $x^* \in X$, that is, $T^r(x^*) = x^*$ for any $r \in \mathbb{N}$

### Exercise C.6 1

Taking inspiration from [I. Altun, B. Damjanovic, and D. Djoric, "Fixed point and common fixed point theorems on ordered cone metric spaces," Applied Mathematics Letters, vol. 23, no. 3, pp. 310–316, 2010] we make the following: Let $(X, \sqsubseteq)$ be a partially ordered set. Two mappings $f, g : X \mapsto X$ will be called $r$-weakly increasing if $fx \sqsubseteq_r gfx$, that is, $f^r g^r x \sqsubseteq g^r f^r x$ and $gx \sqsubseteq_r fgx$.

Using this, prove the following: Let $(X, \sqsubseteq)$ be a partially ordered set and suppose there exists a cone $b$-metric $d$ in $X$ such that $(X, d)$ is a complete cone $b$-metric space with coefficient $s \geq 1$ relative to a solid cone $P$. Let $f, g : X \mapsto X$ be two weakly $r$-increasing mappings with respect to "$\sqsubseteq$". Suppose the following hold

(a) there exist $Z \geq 1$ and $\beta \in [0, \frac{6}{5(s^2+3s+2)})$ modified from Proposition 1.3 [Ampadu, Clement(2016):Higher Order Hardy-Rogers Contraction Mapping Theorem

(b) $d(T^r x, T^r y) \leq Z\beta^r [d(x, Tx) + d(y, Ty) + d(x, Ty) + d(y, Tx) + d(x, y)]$, for all comparative $x, y \in X$

(c) $f$ or $g$ is $r$-continuous, that is, for any $r \in \mathbb{N}$, $f^r$ or $g^r$ is continuous

Then $f$ and $g$ have a common $r$-fixed point $x^* \in X$

### Exercise C.7 1

Let $X$, $E$, $\|u\|$, $P$, and $d : X \times X \mapsto E$ be defined as in Example 18[Abusalim et.al, Fixed Point and Common Fixed Point Theorems on Ordered Cone b-Metric Spaces, Abstract and Applied Analysis, Volume 2013, Article ID 815289, 7 pages, http://dx.doi.org/10.1155/2013/815289]. Then $(X, d)$ is a cone $b$-metric space with coefficient $s = 2$. For any $r \in \mathbb{N}$, define $f : X \mapsto X$ by $f(x) = \frac{x}{2^r}$, then $f$ is $r$-continuous and nondecreasing with respect to "$\sqsubseteq$". Deduce that $f$ is a (modified) ordered higher-order Hardy-Rogers type contraction. Moreover, 0 is the $r$-fixed point of $f$ and all the conditions of the theorem in Exercise C.5 hold.

## 3.5 References

(1) Huang Long-Guang and Zhang Xian, Cone metric spaces and fixed point theorems of contractive mappings, J. Math. Anal. Appl. 332 (2007)1468–1476

(2) G. Jungck, S. Radenovic, S. Radojevıc, and V. Rakocevıc,"Common fixed point theorems for weakly compatible pairs on cone metric spaces," Fixed Point Theory and Applications, vol. 2009, Article ID 643840, pp. 1–13, 2009

(3) A. Azam, M. Arshad, and I. Beg, "Banach contraction principle on cone rectangular metric spaces," Applicable Analysis and Discrete Mathematics, vol. 3, no. 2, pp. 236–241, 2009

(4) S.K. Malhotra et.al, The Scientific World Journal, Volume 2013, Article ID 810732, 7 pages

(5) Ampadu, Clement(2016):Higher Order Hardy-Rogers Contraction Mapping Theorem. Unpublished

(6) S.K. Malhotra et.al, SOME FIXED POINT THEOREMS FOR ORDERED REICH TYPE CONTRACTIONS IN CONE RECTANGULAR METRIC SPACES, Acta Math. Univ. Comenianae Vol. LXXXII, 2 (2013), pp.165-175

(7) Abusalim et.al, Fixed Point and Common Fixed Point Theorems on Ordered Cone b-Metric Spaces, Abstract and Applied Analysis, Volume 2013, Article ID 815289, 7 pages

(8) I. Altun, B. Damjanovic, and D. Djoric, "Fixed point and common fixed point theorems on ordered cone metric spaces," Applied Mathematics Letters, vol. 23, no. 3, pp. 310–316, 2010

# Chapter 4

# Higher-Order Hardy-Rogers Type Graphic Contraction on Cone Rectangular Metric Spaces

## 4.1 Brief Summary

> **Abstract D.1 1**
>
> In this chapter we prove existence and uniqueness of fixed point theorems for a $G$-higher-order Hardy-Rogers type contraction in cone rectangular metric spaces with a graph. We note that the results of this chapter generalize the ordered version of some results in the previous chapter.

## 4.2 Preliminaries

> **Notation D.1 1**
>
> $E$ will denote a real Banach space

> **Notation D.2 1**
>
> $P$ will denote a subset of $E$

> **Definition D.3 1**
>
> [Huang Long-Guang and Zhang Xian, Cone metric spaces and fixed point theorems of contractive mappings, J. Math. Anal. Appl. 332 (2007) 1468–1476] $P$ is called a cone if and only if
>
> ($a$) $P$ is closed, nonempty, and $P \neq \{0\}$
>
> ($b$) $a, b \in \mathbb{R}$, $a, b \geq 0$, $x, y \in P \Rightarrow ax + by \in P$
>
> ($c$) $x \in P$ and $-x \in P \Rightarrow x = 0$

> **Definition D.4 1**
>
> Given cone $P \subset E$, we define a partial ordering $\leq$ with respect to $P$ by $x \leq y$ iff $y - x \in P$

> **Notation D.5 1**
> 
> $x < y$ will mean $x \leq y$, but $x \neq y$, while $x \ll y$ will stand for $y - x \in int\, P$, were $int\, P$ denotes the interior of $P$

> **Definition D.6 1**
> 
> [Huang Long-Guang and Zhang Xian, Cone metric spaces and fixed point theorems of contractive mappings, J. Math. Anal. Appl. 332 (2007) 1468–1476] The cone $P$ is called normal if there is a number $K > 0$ such that for all $x, y \in E$, $0 \leq x \leq y$ implies $\|x\| \leq K\|y\|$. The least positive number satisfying, $0 \leq x \leq y$ implies $\|x\| \leq K\|y\|$, is called the normal constant of $P$

**Definition D.7 1**

[Huang Long-Guang and Zhang Xian, Cone metric spaces and fixed point theorems of contractive mappings, J. Math. Anal. Appl. 332 (2007) 1468–1476] Let $X$ be a nonempty set. Suppose the mapping $d : X \times X \mapsto E$ satisfies

(a) $0 < d(x, y)$ for all $x, y \in X$ and $d(x, y) = 0$ iff $x = y$

(b) $d(x, y) = d(y, x)$ for all $x, y \in X$

(c) $d(x, y) \leq d(x, z) + d(z, y)$ for all $x, y, z \in X$

Then $d$ is called a cone metric on $X$, and $(X, d)$ is called a cone metric space

> **Remark D.8 1**
> 
> We assume $int\, P \neq \emptyset$, that is, $P$ is a solid cone

In the sequel we will need the following

> **Remark D.9 1**
> 
> [G. Jungck, S. Radenovic, S. Radojevic, and V. Rakocevıc, "Common fixed point theorems for weakly compatible pairs on cone metric spaces," Fixed Point Theory and Applications, vol. 2009, Article ID 643840, pp. 1–13, 2009] Let $P \subseteq E$ be a cone, and $a, b, c \in P$, then we have the following
> 
> (a) If $a \leq b$ and $b \ll c$, then $a \ll c$
> 
> (b) If $a \ll b$ and $b \ll c$, then $a \ll c$
> 
> (c) If $0 \leq u \ll c$ for each $c \in int\, P$, then $u = 0$
> 
> (d) If $c \in int\, P$ and $a_n \to 0$, then there exists $n_0 \in \mathbb{N}$ such that, for all $n > n_0$, we have $a_n \ll c$
> 
> (e) If $0 \leq a_n \leq b_n$ for each $n$ and $a_n \to a$, $b_n \to b$, then $a \leq b$
> 
> (f) If $a \leq \lambda a$, where $0 \leq \lambda < 1$, then $a = 0$

# CHAPTER 4. HIGHER-ORDER HARDY-ROGERS TYPE GRAPHIC CONTRACTION ON CONE RECTANGULAR METRIC SPACES

### Definition D.10 1

[A. Azam, M. Arshad, and I. Beg, "Banach contraction principle on cone rectangular metric spaces," Applicable Analysis and Discrete Mathematics, vol. 3, no. 2, pp. 236–241, 2009] Let $X$ be a nonempty set. A mapping $d : X \times X \mapsto E$ is a called a cone rectangular metric on $X$ if it satisfies the following

(a) $0 \leq d(x,y)$ for all $x,y \in X$ and $d(x,y) = 0$ iff $x = y$

(b) $d(x,y) = d(y,x)$ for all $x,y \in X$

(c) $d(x,y) \leq d(x,z) + d(z,y)$ for all $x,y,z \in X$

(d) $d(x,y) \leq d(x,w) + d(w,z) + d(z,y)$ for all $x,y \in X$ and for all distinct points $w,z \in X$-$\{x,y\}$

We say $(X,d)$ is a cone rectangular metric space

### Definition D.11 1

Let $\{x_n\}$ be a sequence in $(X,d)$ and $x \in X$. If for every $c \in E$, with $0 \ll c$ there is a natural number $n_0$ such that for all $n > n_0$, $d(x_n, x) \ll c$, then $\{x_n\}$ is said to be convergent to $x \in X$

### Definition D.12 1

If for every $c \in E$ with $0 \ll c$ there is a natural number $n_0$ such that for all $n > n_0$ and a natural number $m$, we have $d(x_n, x_m) \ll c$, then $\{x_n\}$ is called a Cauchy sequence in $(X,d)$

### Definition D.13 1

If every Cauchy sequence is convergent in $(X,d)$, then $(X,d)$ is called a complete cone rectangular metric space

### Remark D.14 1

If the underlying cone is normal then $(X,d)$ is called a normal cone rectangular metric space

### Example D.15 1

[Satish Shukla, Reich Type Contractions on Cone Rectangular Metric Spaces Endowed with a Graph,Theory and Applications of Mathematics and Computer Science 4 (1) (2014) 14–25] Let $X = \mathbb{N}$, $E = \mathbb{R}^2$, $\alpha, \beta > 0$ and $P = \{(x,y) : x,y \geq 0\}$. Define $d : X \times X \mapsto E$ as follows: $d(x,y) = (0,0)$ if $x = y$; $d(x,y) = 3(\alpha, \beta)$ if $x$ and $y$ are in $\{1,2\}$, $x \neq y$; $d(x,y) = (\alpha, \beta)$, otherwise. Then $(X,d)$ is a cone rectangular metric space but not a cone metric space

### Example D.16 1

[Satish Shukla, Reich Type Contractions on Cone Rectangular Metric Spaces Endowed with a Graph,Theory and Applications of Mathematics and Computer Science 4 (1) (2014) 14–25] Let $X = \mathbb{N}$, $E = C^1_{\mathbb{R}}[0,1]$ with $\|x\| = \|x\|_\infty + \|x'\|_\infty$ and $P = \{x \in E : x(t) \geq 0 \text{ for } t \in [0,1]\}$. Then this cone is not normal. However, defining $d : X \times X \mapsto E$ as follows: $d(x,y) = 0$ if $x = y$; $d(x,y) = 3e^t$ if $x,y \in \{1,2\}$, $x \neq y$; $d(x,y) = e^t$, otherwise. Then $(X,d)$ is a non-normal cone rectangular metric space but $(X,d)$ is not a cone metric space.

# CHAPTER 4. HIGHER-ORDER HARDY-ROGERS TYPE GRAPHIC CONTRACTION ON CONE RECTANGULAR METRIC SPACES

**Notation D.17 1**

Let $(X, d)$ be a metric space. $\triangle$ will denote the diagonal of the Cartesian product $X \times X$

**Notation D.18 1**

Let $G$ be a directed graph. $V(G)$ will denote the set of vertices of the graph and $E(G)$ will denote the set of edges of the graph.

**Remark D.19 1**

We assume $V(G)$ coincides with $X$. We also assume that $E(G)$ contains all loops, that is, $E(G) \supseteq \triangle$.

**Remark D.20 1**

If $G$ has no parallel edges, one can identify $G$ with the pair $(V(G), E(G))$

**Definition D.21 1**

The conversion of the graph $G$ is denoted by $G^{-1}$, which is a graph obtained from $G$ by reversing the direction of the edges. In particular, $E(G^{-1}) = \{(y, x) \in X \times X : (x, y) \in E(G)\}$

**Notation D.22 1**

$\widetilde{G}$ will denote the undirected graph obtained from $G$ by omitting the direction of the edges.

**Remark D.23 1**

If $\widetilde{G}$ is a directed graph for which $E(\widetilde{G})$ is symmetric, then $E(\widetilde{G}) = E(G) \cup E(G^{-1})$

**Definition D.24 1**

If $x$ and $y$ are vertices in a graph $G$, then a path from $x$ to $y$ of length $N \in \mathbb{N}$ is a sequence $\{x_i\}_{i=0}^{N}$ of $N+1$ vertices such that $x_0 = x$, $x_N = y$ and $(x_{n-1}, x_n) \in E(G)$ for $i = 1, \cdots, N$

**Definition D.25 1**

Let $(X, d)$ be a cone rectangular metric space endowed with a graph $G$. A mapping $T : X \mapsto X$ will be called a $G$-Higher-Order-Hardy-Rogers-Type contraction if the following conditions hold

(a) $T$, $r$-preserves edges of $G$, that is, $(x, y) \in E(G) \Rightarrow (T^r x, T^r y) \in E(G)$ for all $x, y \in X$ and any $r \in \mathbb{N}$

(b) $T$, $r$-decreases weights of edges of $G$, that is, $(x, y) \in E(G) \Rightarrow d(T^r x, T^r y) \leq Z\beta^r [d(x, Tx) + d(y, Ty) + d(x, Ty) + d(y, Tx) + d(x, y)]$ for all $x, y \in X$ and any $r \in \mathbb{N}$, where $Z \geq 1$ and $\beta \in [0, \frac{1}{5})$ are given by Proposition 1.3[Ampadu, Clement(2016):Higher Order Hardy-Rogers Contraction Mapping Theorem. Unpublished]

**Remark D.26 1**

If $T$ is a $G$-higher-order-Hardy-Rogers-type contraction then it is both a $G^{-1}$-higher-order Hardy-Rogers-type contraction and a $\widetilde{G}$-higher-order-Hardy-Rogers-type contraction

### Example D.27 1

Define $T : X \mapsto X$ for any $r \in \mathbb{N}$ by $T^r x = c$, where $c \in X$ is fixed, then $T$ is a $G$-higher-order-Hardy-Rogers-type contraction since $E(G)$ contains all loops

### Example D.28 1

Any higher-order-Hardy-Rogers-type contraction on a set $X$ is a $G_0$-higher-order-Hardy-Rogers-type contraction, where $E(G_0) = X \times X$

Taking inspiration from [Malhotra, S. K., S. Shukla and R. Sen (2013b). Some fixed point theorems for ordered reich type contractions in cone rectangular metric spaces. Acta Mathematica Universitatis Comenianae (2), 165–175] we have the following

### Example D.29 1

Let $(X, d)$ be a cone rectangular metric space, and "$\sqsubseteq$" be a partial order on $X$ and $T : X \mapsto X$ be an ordered higher-order Hardy-Rogers type contraction, that is, $T$ satisfies, $d(T^r x, T^r y) \leq Z\beta^r [d(x, Tx) + d(y, Ty) + d(x, Ty) + d(y, Tx) + d(x, y)]$ for all $x, y \in X$ and any $r \in \mathbb{N}$, where $Z \geq 1$ and $\beta \in [0, \frac{1}{5})$ are given by Proposition 1.3[Ampadu, Clement(2016):Higher Order Hardy-Rogers Contraction Mapping Theorem.Unpublished], with $x \sqsubseteq y$. Then $T$ is a $G_1$-higher-order Hardy-Rogers type contraction, where $E(G_1) = \{(x, y) \in X \times X : x \sqsubseteq y\}$

Recall that higher-order contraction mapping principle was given in [Ezearn Fixed Point Theory and Applications (2015) 2015:88] and an alternate characterization was given in [Ampadu, Clement (2015): Generalization of Higher Order Contraction Mapping Theorem. Unpublished]. It follows from these papers and the above remark, that we have the following.

### Definition D.30 1

A map $f$ from a metric space $(X, d)$ into itself will be called a weakly $r$-Picard operator ($r$-WPO) if $\lim_{n \to \infty} f^{rn} x = z$ for all $x \in X$ and any $r \in \mathbb{N}$, and $z$ is a $r$-fixed point of $f$, that is, $f^r z = z$ for any $r \in \mathbb{N}$. Moreover, the $r$-fixed point of $f$ is unique, and we say that $f$ is a $r$-Picard operator ($r$-PO).

### Remark D.31 1

Throughout we assume that $G$ is a directed graph such that $V(G) = X$ and $E(G) \supseteq \Delta$

### Definition D.32 1

The set of all $r$-fixed points of the mapping $T : X \mapsto X$ will be denoted by $Fix_r(T)$, that is, $Fix_r(T) = \{x \in X : T^r x = x\}$

### Definition D.33 1

The set of all $r$-periodic points of $T$ will be denoted by $P_r(T)$, that is, for some $n \in \mathbb{N}$ and any $r \in \mathbb{N}$, $P_r(T) = \{x \in X : T^{rn}(x) = x\}$

### Definition D.34 1

$X_T^r$ will denote the set $\{x \in X : (x, T^r x), (x, T^{2r} x) \in E(G)\}$

# CHAPTER 4. HIGHER-ORDER HARDY-ROGERS TYPE GRAPHIC CONTRACTION ON CONE RECTANGULAR METRIC SPACES

**Definition D.35 1**

We will say $(X,d)$ have the property $P$ if whenever $\{x_n\}$ is a sequence in $X$ converging to $x$ with $(x_n, x_{n+1}) \in E(G)$ for all $n \in \mathbb{N}$, there is a subsequence $\{x_{n_k}\}$ with $(x_{n_k}, x) \in E(G)$ for all $n \in \mathbb{N}$

## 4.3 Main Results

**Proposition D.1 1**

Let $(X,d)$ be a cone rectangular metric space endowed with a graph $G$ and $T: X \mapsto X$ be a $G$-higher-order-Hardy-Rogers-type contraction. If $x, y \in Fix_r(T)$ are such that $(x,y) \in E(G)$, then $x = y$.

**Proof of Proposition D.1 1**

Let $x, y \in Fix_r(T)$ and $(x,y) \in E(G)$, then since $T: X \mapsto X$ is a $G$-higher-order-Hardy-Rogers-type contraction, we deduce that,

$$d(x,y) = d(T^r x, T^r y)$$
$$\leq 3Z\beta^r d(x,y)$$

However, $1 - 3Z\beta^r > 0$, thus, $d(x,y) = 0$, that is, $x = y$

**Theorem D.2 1**

Let $(X,d)$ be a cone rectangular metric space endowed with a graph $G$. Let $T: X \mapsto X$ be a $G$-higher-order-Hardy-Rogers-type contraction. Then for every $x_0 \in X_T^r$, the $r$-Picard sequence $\{x_n\}$ is a Cauchy sequence.

> **Proof of Theorem D.2 1**
>
> Let $x_0 \in X_T^r$ and define the iterative sequence $\{x_n\}$ by $x_{n+1} = T^r x_n$ for all $n \geq 0$. Since $x_0 \in X_T^r$ we have $(x_0, T^r x_0) \in E(G)$ and $T$ is a $G$-higher-order Hardy-Rogers type contraction. Furthermore we have $(T^r x_0, T^{2r} x_0) = (x_1, x_2) \in E(G)$. By induction we obtain $(x_n, x_{n+1}) \in E(G)$ for all $n \geq 0$. Now since $(x_n, x_{n+1}) \in E(G)$ for all $n \geq 0$ and $T$ is a $G$-higher-order Hardy-Rogers type contraction, we have,
>
> $$\begin{aligned} d(x_n, x_{n+1}) &= d(T^r x_{n-1}, T^r x_n) \\ &\leq Z\beta^r [d(x_{n-1}, x_n) + d(x_{n-1}, Tx_{n-1}) + d(x_n, Tx_n) \\ &\quad + d(x_{n-1}, Tx_n) + d(x_n, Tx_{n-1})] \\ &= Z\beta^r [d(x_n, x_{n+1}) + d(x_{n-1}, x_n) + d(x_{n-1}, x_{n+1}) + d(x_n, x_{n-1})] \\ &\leq Z\beta^r [2d(x_n, x_{n+1}) + 3d(x_{n-1}, x_n)] \end{aligned}$$
>
> From the above it follows that $d(x_{n+1}, x_n) \leq k d(x_n, x_{n-1})$, where $k := \frac{3Z\beta^r}{1-2Z\beta^r}$. Set $d_n = d(x_n, x_{n+1})$ for all $n \geq 0$, then by induction we obtain that $d_n \leq k^n d_0$ for all $n \in \mathbb{N}$. Note that if $x_0 \in P_r(T)$, then there exists $q \in \mathbb{N}$ such that $T^{qr} x_0 = x_q = x_0$ and since $d_n \leq k^n d_0$, we have, $d_0 = d(x_0, x_1) = d(x_0, T^r x_0) = d(x_q, T^r x_q) = d(x_q, x_{q+1}) \leq k^q d(x_0, x_1) = k^q d_0$. Since $k \in [0, 1)$, we get a contradiction. Thus, we assume that $x_n \neq x_m$ for all distinct $n, m \in \mathbb{N}$. As $x_0 \in X_T^r$, we have, $(x_0, T^{2r} x_0) = (x_0, x_2) \in E(G)$ and since $T$ is a $G$-higher-order Hardy-Rogers type contraction, we obtain $(T^r x_0, T^r x_2) = (x_1, x_3) \in E(G)$. By induction we obtain $(x_n, x_{n+2}) \in E(G)$ for all $n \geq 0$. Since $T$ is a $G$-higher-order Hardy-Rogers type contraction, it follows that,
>
> $$\begin{aligned} d(x_n, x_{n+2}) &\leq k d_{n-1} + k d_{n+1} \\ &\leq k^n d_0 + k^{n+2} d_0 \\ &= (1 + k^2) k^n d_0 \end{aligned}$$
>
> Let $\beta = 1 + k^2 \geq 0$, then $d(x_n, x_{n+2}) \leq \beta k^n d_0$. Now for the sequence $\{x_n\}$, we consider $d(x_n, x_{n+p})$ in two cases. If $p$ is odd say $2m+1$, then we deduce that $d(x_n, x_{n+2m+1}) \leq \frac{k^n}{1-k} d_0$. If $p$ is even, say $2m$, then we deduce that $d(x_n, x_{n+2m}) \leq \frac{k^n}{1-k} d_0 + \beta k^n d_0$. As $\beta \geq 0$ and $0 \leq k < 1$, we have, $\frac{k^n}{1-k} d_0 \to 0$ and $\beta k^n d_0 \to 0$. Thus, it follows that for every $c \in E$ with $0 \ll c$ there exists a natural number $n_0$ such that $d(x_n, x_{n+2m+1}) \ll c$ and $d(x_n, x_{n+2m}) \ll c$ for all $n > n_0$. Thus $\{x_n\}$ is a Cauchy sequence

> **Theorem D.3 1**
>
> Let $(X, d)$ be a complete cone rectangular metric space endowed with a graph G and has the property (P). Let $T : X \mapsto X$ be a G-higher-order Hardy-Rogers type contraction such that $X_T^r \neq \emptyset$, then $T$ is a weakly $r$-Picard operator.

### Proof of Theorem D.3 1

If $X_T^r \neq \emptyset$, then let $x_0 \in X_T^r$. By previous theorem, the $r$-Picard sequence $\{x_n\}$, where $x_n = T^{r(n-1)}$ for all $n \in \mathbb{N}$ is a Cauchy sequence in $X$. Since $X$ is complete there exists $u \in X$ such that $x_n \to u$ as $n \to \infty$. We show $u$ is a $r$-fixed point of $T$. By the previous theorem, we have $(x_n, x_{n+1}) \in E(G)$ for all $n \geq 0$, $d_n \leq d(x_n, x_{n+1}) \leq k^n d_0$, and by the property (P) there exists a subsequence $\{x_{n_v}\}$ such that $(x_{n_v}, u) \in E(G)$ for all $n \in \mathbb{N}$. Let $x_n \neq x_{n-1}$ for all $n \in \mathbb{N}$. Now,

$$d(u, T^r u) \leq d(u, x_{n_v}) + d(x_{n_v}, x_{n_v+1}) + d(x_{n_v+1}, T^r u)$$
$$= d(u, x_{n_v}) + d_{n_v} + d(T^r x_{n_v}, T^r u)$$
$$\leq d(u, x_{n_v}) + d_{n_v} + Z\beta^r [2d(u, T^r u) + 2d(u, x_{n_v}) + d_{n_v}]$$

From the above one deduces that

$$d(u, T^r u) \leq \frac{1 + 2Z\beta^r}{1 - 2Z\beta^r} d(u, x_{n_v}) + \frac{1 + 2Z\beta^r}{1 - 2Z\beta^r} d_{n_v}$$
$$\leq \frac{1 + 2Z\beta^r}{1 - 2Z\beta^r} d(u, x_{n_v}) + \frac{1 + 2Z\beta^r}{1 - 2Z\beta^r} k^{n_v} d_0$$

Since $k^{n_v} d_0 \to 0$, $x_n \to u$ as $n \to \infty$, we can choose $n_0 \in \mathbb{N}$ such that for every $c \in E$ with $0 \ll c$, we have, $d(u, x_{n_v}) \ll \frac{c(1-2Z\beta^r)}{2(1+2Z\beta^r)}$ and $k^{n_v} d_0 \ll \frac{c(1-2Z\beta^r)}{2(1+2Z\beta^r)}$. Consequently, for every $c \in E$ with $0 \ll c$ we have $d(u, T^r u) \ll c$ for all $n > n_0$. Consequently, we have $d(u, T^r u) = 0$, that is, $u = T^r u$, therefore $u \in Fix_r(T)$. Thus $T$ is a weakly $r$-Picard operator.

### Theorem D.4 1

Let $(X, d)$ be a complete cone rectangular metric space endowed with a graph $G$ and has the property $P$. Let $T : X \mapsto X$ be a $G$-higher-order Hardy-Rogers type contraction such that $X_T^r \neq \emptyset$, then $T$ is a weakly $r$-Picard operator. Furthermore the subgraph $G_{Fix_r}$ defined by $V(G_{Fix_r}) = Fix_r T$ is $r$-weakly connected iff $T$ is a $r$-Picard operator.

### Proof of Theorem D.4 1

The existence of fixed points follows from the previous theorem. Let $u, v \in Fix_r T$, then since $G_{Fix_r}$ is $r$-weakly connected there exists a path $(x_i)_{i=0}^{l}$ in $G_{Fix_r}$ from $u$ to $v$, that is, $x_0 = u$, $x_l = v$ and $(x_{i-1}, x_i) \in E(G_{Fix_r})$ for $i = 1, 2, \cdots, l$, therefore by Proposition D.1 and Remark D.9 we obtain $u = v$. Thus, the fixed point is unique and $T$ is a $r$-Picard operator

### Corollary D.5 1

Let $(X, \sqsubseteq, d)$ be an ordered cone rectangular metric space, $f : X \mapsto X$ be a mapping such that the following conditions are satisfied

(a) $f$ is an ordered higher-order Hardy-Rogers type contraction

(b) $f$ is $r$-nondecreasing with respect to "$\sqsubseteq$"

(c) there exists $x_0 \in X$ such that $x_0 \sqsubseteq f^r x_0$

(d) if $\{x_n\}$ is a nondecreasing sequence in $X$ converging to some $z$, then $x_n \sqsubseteq z$ for all $n$

Then, $f$ is a $r$-weakly Picard operator. In addition, $Fix_r(f)$ is well-ordered if and only if $f$ is a $r$-Picard operator

> **Proof of Corollary D.5 1**
>
> Let $G$ be a graph defined by $V(G) = X$ and $E(G) = \{(x,y) \in X \times X : x \sqsubseteq y\}$. Then by (a) and (b), $f$ is a $G$-higher-order Hardy Rogers type contraction and by (c) we have $X_T^r \neq \emptyset$. Also by (d), property (P) is satisfied. The result follows from the previous theorem.

## 4.4 Exercises

> **Exercise D.1 1**
>
> Let $(X, \sqsubseteq, d)$ be an ordered cone rectangular metric space and $f, g : X \mapsto X$ be two mappings. Let $f$ be an ordered $r$-$g$-weak higher-order Hardy-Rogers type contraction, that is, $d(f^r x, f^r y) \leq Z\beta^r [d(g^r x, fx) + d(g^r y, fy) + d(g^r x, fy) + d(g^r y, fx) + d(g^r x, g^r y)]$, for all $x, y \in X$ with $gx \sqsubseteq_r gy$, where $Z \geq 1$ and $\beta \in [0, \frac{1}{5})$ come from Proposition 1.3 [Ampadu, Clement(2016):Higher Order Hardy-Rogers Contraction Mapping Theorem. Unpublished]. Deduce that $f$ is a $G_2$-higher-order-Hardy-Rogers type contraction, where $E(G_2) = \{(x, y) \in X \times X : gx \sqsubseteq_r gy\}$

> **Exercise D.2 1**
>
> Let $(X, d)$ be a cone metric space and $T, S : X \mapsto X$ be two functions. A mapping $S$ is said to be a $T$-Reich contraction[Sandeep Bhatt, Amit singh and R.C. Dimri, Fixed point theorems for certain contractive Mappings in cone metric spaces, Int. journal of math. Archive-2(4), (2011), 444 – 45] if it satisfies $d(TSx, TSy) \leq ad(Tx, TSx) + bd(Ty, TSy) + cd(Tx, Ty)$ for all $x, y \in X$, where $a, b, c \geq 0$ and satisfy $a + b + c < 1$. Deduce there exist $M \geq 1$ and $\zeta \in [0, \frac{1}{3})$ obtained in a similar manner as in Proposition 1.3 [Ampadu, Clement(2016):Higher Order Hardy-Rogers Contraction Mapping Theorem. Unpublished] such that $d(TS^r x, TS^r y) \leq M\zeta^r [d(Tx, TSx) + d(Ty, TSy) + d(Tx, Ty)]$, thus we have a characterization of what it means for a mapping $S$ to be higher-order $T$-Reich type contraction mapping.

> **Exercise D.3 1**
>
> Deduce that any higher-order $T$-Reich type contraction on a set $X$ is a $G_0$-higher-order-$T$-Reich type contraction, where $E(G_0) = X \times X$

> **Exercise D.4 1**
>
> Define $S, T : X \mapsto X$ by $TS^r x = Tc$, for any $r \in \mathbb{N}$, where $c \in X$ is fixed. Deduce that $S$ is a $G$-higher-order-$T$-Reich type contraction

> **Exercise D.5 1**
>
> Let $(X, d)$ be a cone rectangular metric space endowed with a graph $G$ and $S : X \mapsto X$ be a $G$-higher-order-$T$-Reich type contraction. If $Tx, Ty \in Fix_r(S) = \{Tx \in X : TS^r x = Tx\}$ are such that $(Tx, Ty) \in E(G)$, then $Tx = Ty$.

> **Exercise D.6 1**
>
> Let $S, T : X \mapsto X$ be such that $S$ is $G$-higher-order-$T$-Reich type contraction
>
> (a) Give a characterization of Theorem D.2 and prove it
>
> (b) Give a characterization of Theorem D.3 and prove it
>
> (c) Give a characterization of Theorem D.4 and prove it
>
> (d) Give a characterization of Corollary D.5 and prove it

## 4.5 References

(1) Huang Long-Guang and Zhang Xian, Cone metric spaces and fixed point theorems of contractive mappings, J. Math. Anal. Appl. 332 (2007) 1468–1476

(2) G. Jungck, S. Radenovic, S. Radojevıc, and V. Rakocevıc,"Common fixed point theorems for weakly compatible pairs on cone metric spaces," Fixed Point Theory and Applications, vol. 2009, Article ID 643840, pp. 1–13, 2009

(3) A. Azam, M. Arshad, and I. Beg, "Banach contraction principle on cone rectangular metric spaces," Applicable Analysis and Discrete Mathematics, vol. 3, no. 2, pp. 236–241, 2009

(4) Satish Shukla, Reich Type Contractions on Cone Rectangular Metric Spaces Endowed with a Graph,Theory and Applications of Mathematics and Computer Science 4 (1) (2014) 14–25

(5) Ampadu, Clement(2016):Higher Order Hardy-Rogers Contraction Mapping Theorem. Unpublished

(6) Malhotra, S. K., S. Shukla and R. Sen (2013b). Some fixed point theorems for ordered reich type contractions in cone rectangular metric spaces. Acta Mathematica Universitatis Comenianae (2), 165–175

(7) Ezearn Fixed Point Theory and Applications (2015) 2015:88

(8) Ampadu, Clement (2015): Generalization of Higher Order Contraction Mapping Theorem. Unpublished

(9) Sandeep Bhatt, Amit singh and R.C. Dimri, Fixed point theorems for certain contractive Mappings in cone metric spaces, Int. journal of math. Archive-2(4), (2011), 444 – 45